Food Safety in China

Weisheng Zhou
Editor

Food Safety in China

The Three-Layer Trust Model and Governance Transformation in the Digital Age

Springer

Editor
Weisheng Zhou
College of Policy Science
Ritsumeikan University
Kyoto, Japan

ISBN 978-981-92-3740-1 ISBN 978-981-92-3741-8 (eBook)
https://doi.org/10.1007/978-981-92-3741-8

This Springer imprint is published by the registered company Springer Nature Singapore Pte Ltd.
The registered company address is: 152 Beach Road, #21-01/04 Gateway East, Singapore 189721, Singapore

Preface

"民以食爲天"

"Food is the foundation of life and society."

As early as approximately two thousand years ago, China's *Book of Han* (*Hanshu*) recorded that food is not only the foundation of human survival, but also the basis for social stability and continuity. Here, the term "food" refers to a concept that encompasses both sufficient quantity and assured quality and safety. **The central message of this book is straightforward yet significant: food safety in contemporary China is not merely a matter of hygiene or quality control, but a problem fundamentally rooted in the structure of social trust.**

Since the reform and opening-up period, China has largely overcome chronic food shortages through improvements in agricultural productivity and the development of distribution networks, thereby providing an answer to the quantitative question of "who will feed China." However, once the stability of supply was achieved, a new qualitative question emerged: how to feed the population safely and sustainably. With a population of over 1.4 billion, China is now confronted simultaneously with the dual challenges of food security and food safety. In particular, food safety has become one of the most pressing public concerns amid rapid economic growth and urbanization. Incidents such as the 2008 melamine-contaminated milk scandal, along with cases involving counterfeit food, pesticide residues, and illegal additives, have not only heightened public anxiety about health risks but also undermined institutional trust in government and enterprises.

Food safety has thus evolved beyond isolated incidents of contamination or regulatory failure into a comprehensive issue that cuts across state governance, corporate ethics, and market institutions. Moreover, with the rise of the digital economy, the nature of food safety challenges has undergone profound transformation. The expansion of complex supply chains, the rapid growth of online food delivery (OFD) markets, and the proliferation of platform-based economies have shifted food safety risks from individual and localized risks to structural and systemic risks. Consumers are increasingly unable to directly observe food production and processing and are therefore placed in a position of reliance on "invisible risks." Under such conditions, safety is no longer determined solely by physical quality but is socially constructed through institutions, information systems, and platform design. As a result, even when objective safety conditions improve, consumer anxiety and distrust do not necessarily diminish. This divergence reflects the interplay of multiple factors, including gaps in risk perception, information asymmetry, and varying levels of institutional trust.

While previous research has primarily focused on scientific risk assessment and regulatory frameworks, relatively little attention has been paid to trust itself as a central analytical concept. This book addresses precisely this gap. It repositions food safety from a problem of risk management to one of trust construction. **Specifically, it conceptualizes food safety through an integrated framework consisting of three interrelated layers: psychological trust (individual level), institutional trust (governance level), and structural trust (societal level).**

Based on this three-layered model of trust, the book develops a step-by-step analysis.

Part I examines risk perception and consumer behavior, elucidating how the gap between subjective risk and objective safety generates distrust.

Part II explores institutional reforms and governance mechanisms, analyzing how trust is formed and maintained at the systemic level.

Part III investigates how the digital economy reshapes the structure of food safety risks, particularly through the expansion of platform-based systems.

The key conclusion of this book is that food safety should not be left to the efforts of individual actors alone, but must be understood as an institutional public good produced through the integration of institutions, markets, technology, and society.

As a classic example of a credence good, food cannot be fully governed by market mechanisms alone. Therefore, the construction of collaborative governance involving the state, enterprises, digital platforms, and consumers is indispensable. Furthermore, food safety is closely linked to the broader agenda of sustainability. It is deeply connected to the Sustainable Development Goals (SDGs), including health (SDG 3), zero hunger (SDG 2), responsible consumption and production (SDG 12), and strong institutions (SDG 16).

In an era of globalization and digitalization, food safety must be reexamined not merely as a domestic policy issue but as a fundamental challenge in the design of social systems. Against this background, this book aims to provide policy implications for rebuilding trust through an in-depth analysis of China's experience, while also presenting a new analytical framework that integrates theory, empirical analysis, institutions, and structural transformation.

The insights developed here are intended not only to advance the study of food safety but also to contribute to the broader field of public policy design under conditions of increasing uncertainty. Finally, it should be noted that the issue of food safety addressed in this volume presupposes the stability of food supply, namely food security. This related dimension will be examined systematically in a separate volume dedicated to food security.

Kyoto, Japan Weisheng Zhou
May 2026

Contents

Editors and Contributors

About the Editor

Weisheng Zhou, Ph.D. in Engineering (Kyoto University), Emeritus Professor at Ritsumeikan University; Qiushi Distinguished Professor at Zhejiang University; Director of the Anhui Provincial Academician Workstation for Sustainable Development and Carbon-neutral Technologies.

Dr. Zhou has held various academic and research positions, including Researcher at the New Energy and Industrial Technology Development Organization (NEDO); Senior Researcher and Research Advisor at the Research Institute of Innovative Technology for the Earth (RITE); Specially Appointed Professor at Osaka University; Visiting Professor at Peking University and Kyoto University; and Founding Director of the Ritsumeikan Research Center for Sustainability Science.

His main fields of teaching and research include Earth Environmental Studies, Environmental Economics and Policy, Energy Systems Engineering, Sustainable Development Studies, and Policy Engineering. His major research interests encompass the East Asian Low-Carbon Community (EA-LCC), the Global Resource Recycling System (GRS), the East Asian Nuclear Power Safety System (EA-NPSS), the East Asian Emissions Trading System (EA-ETS), Quantitative Happiness Studies.

Dr. Zhou has authored and co-authored numerous books, East Asian Low-Carbon Community (Springer, 2021), Sustainable science in the Era of SDGs (Horitsu Bunka Sha, 2022), Nuclear Power Safety and Governance in East Asia (Routledge, 2023), Creep-Fatigue Fracture: Analysis of Internal Damage (Springer, 2024), Foundations of Quantitative Happiness: Economic Growth vs. Life Satisfaction (Springer Nature, 2025), and Decarbonization Policy Engineering: Energy Transition and Regional Cooperation in East Asia (Springer, 2026), among many other publications.

He is a Foreign Fellow of the Engineering Academy of Japan (EAJ).

Contributors

Fengyang Wang, Ph.D., Chief Research Fellow, Research Institute of Global 3E, Visiting Professor at Xi'an Peihua University.

Dr. Wang earned his Ph.D. from the Graduate School of Policy Science at Ritsumeikan University in 2017. He currently serves as a visiting professor at the School of International Education, Xi'an Peihua University. He previously served as Director of Human Resources at two Fortune Global 500 companies, Country Garden Group and Hailiang Group, while also holding a concurrent position as Chief Research Fellow at the Research Institute of Global 3E. His areas of specialization are environmental policy and sustainability studies.

Dr. Wang has addressed a wide range of societal challenges, including food safety, food security, Japan-China relations, regional cooperation in East Asia, resource sustainability, and the realization of a low-carbon society, through interdisciplinary and systems-based approaches. His research focuses on consumer trust in food safety in China, grounded in discussions on the current situation surrounding food safety in the country, and provides policy recommendations based on both theoretical and empirical studies.

During his doctoral studies, he received several honors, including the Kobe Mayor's Award and the Excellent Paper Award from the Japan-China Relations Academic Society.

Xuepeng Qian, Ph.D., Professor of Graduate School of Global Environmental Studies, Director of Institute for Studies of the Global Environment, Sophia University.

Dr. Qian received his Ph.D. in Environmental Systems Engineering from Ritsumeikan University in 2007. He subsequently worked as a lecturer in the same department and later served as an assistant professor and associate professor at Ritsumeikan Asia Pacific University from 2010 to 2021 before assuming his current position.

His research lies in the field of urban sustainability science, with expertise in urban environmental planning. He has worked on a broad range of urban and environmental issues, including urban development, transportation, resource sustainability, water and energy systems, and low-carbon transitions. Using interdisciplinary and systems-based approaches, his work explores how urban and regional systems can be guided toward more sustainable, resilient, and low-carbon development pathways.

His recent research examines sustainable behavior, urban social systems, regional circular systems, industrial decarbonization, low-carbon mobility, and the synergies between decarbonization and city revitalization. His work aims to contribute to interdisciplinary methodologies and practical strategies for sustainable urban and regional transformation.

Dr. Qian has published more than 120 academic works, including journal articles, book chapters, and books, in English and Japanese. He has also served as an editor and guest editor for several international journals.

Part I
Risk Perception and Consumer Behavior in Food Safety: Mechanisms of Psychological Trust Formation

Chapter 1
The Risk Structure and Institutional Challenges of Food Safety in China: The Problematic of Qualitative Safety and Trust

Weisheng Zhou

1.1 The Issue at Hand the Contemporary Challenge of Food Qualitative Safety (Food Safety)

The Chinese idiom "*min yi shi wei tian*" concisely captures the ancient Chinese philosophy that "food is of primary importance to the people". Food is not merely a source of nutrition, it is also a vital element that underpins the stability of people's lives, the maintenance of social order, and even the legitimacy of state governance. This traditional value system continues to exert a powerful influence in modern Chinese society. However, the rapid economic growth and progress of marketisation since the Reform and Opening-up have significantly transformed the structure of food production, distribution and consumption, giving rise to new challenges that cannot be fully addressed by the conventional concept of "food hygiene". Against this backdrop, this book focuses on the issue of food "qualitative safety" and aims to clarify how consumer trust is formed and what factors determine it.

1.1.1 The Dual Structure of China's Food Problem: Food Security (Quantitative Safety) and Food Safety (Qualitative Safety)

When discussing food safety issues in China, the first thing to note is the dual structure comprising "quantitative safety" and "qualitative safety".

W. Zhou (✉)
College of Policy Science, Ritsumeikan University, Kyoto, Japan
e-mail: zhou@sps.ritsumei.ac.jp

W. Zhou (ed.), *Food Safety in China*, https://doi.org/10.1007/978-981-92-3741-8_1

1.1.1.1 Quantitative Safety (Food Security)

Food security is defined as "a state in which all people have physical and economic access to sufficient, safe and nutritious food at all times" (EC-FAO Food Security Programme, 2008). The question is whether China, with a population of 1.4 billion, can ensure a stable supply of the necessary calories and basic agricultural products; this issue relates to grain self-sufficiency rates, production capacity, dependence on imports and stockpiling systems. It was the American environmentalist Lester R. Brown who first raised this issue on the international stage. In his book *Who Will Feed China?*, he warned that, against a backdrop of China's population growth, shrinking arable land, water resource constraints and advancing urbanization, China's importation of large quantities of grain could strain international markets and potentially trigger a global food crisis (Brown, 1995).

Consequently, thanks to agricultural reforms and marketisation in China, as well as the introduction of agricultural technology, food production capacity has improved significantly, and the grain self-sufficiency rate has remained at a high level. As a result, extreme hunger has decreased substantially, and the situation is assessed as being institutionally manageable in terms of food security. Viewed in the light of the United Nations Sustainable Development Goals (SDGs) 2 "End hunger, achieve food security and improved nutrition, and promote sustainable agriculture", these developments can be seen as an example of significant progress towards achieving the goal. In this sense, the crisis envisaged by Brown has not yet materialized. Furthermore, having peaked in 2021, China's total population has entered a phase of decline, and its demographic trends are undergoing a structural transformation. This could serve as a factor that mitigates future growth in food demand to a certain extent. Taking these circumstances into account, it is possible to assess that China has provided a certain institutional response to the fundamental question of "who will feed the Chinese people?", a question that has persisted since the late twentieth century.

However, in a country with a population of 1.4 billion, the surge in demand for feed grains and soya beans accompanying the expansion of meat consumption has increased reliance on imports, thereby linking food supply and demand closely to the international market (OECD & FAO, 2023). Constraints on arable land, regional imbalances in water resources and the risks of climate change also remain, and the "pressure of quantity" continues, albeit in a different form. Quantitative security has evolved into a complex security issue encompassing not only domestic production capacity but also international division of labor and geopolitical stability. Quantitative food security remains the cornerstone of China's food policy.

1.1.1.2 Qualitative Safety (Food Safety)

On the other hand, food safety is defined as "a state in which food does not cause harm to health at the point of consumption" (World Health Organization, 2015). Food safety is not merely a matter of food safety and nutritional value, but also concerns consumer trust. With economic development, urbanization, rising income levels and the expansion of the middle class, public concern has shifted from "how much can be eaten" to a focus on qualitative aspects, namely "how can food be consumed safely and healthily".

In particular, risks relating to food safety, such as pesticide residues, heavy metal contamination, microbial hazards, hygiene management during processing and distribution, and imbalances in supervisory systems, are dispersed throughout the entire supply chain, from production to consumption. In particular, the 2008 Sanlu infant formula scandal, discussed in Sect. 1.1.2.2, had a profound impact on Chinese society and made it clear that "safety" and "trustworthiness" are at the heart of food issues. Since this incident, amendments to food safety legislation and the reorganization of supervisory systems have been implemented; however, the increasing sophistication of processed foods, the expansion of the catering and delivery markets, and the lengthening of supply chains have dispersed the points of risk origin, further complicating the issues.

Qualitative food safety is evolving into a structural governance issue encompassing multiple dimensions, including not only regulatory frameworks but also corporate ethics, transparency, traceability, consumer participation, environmental impact and sustainability. Consequently, China's food safety issue does not involve a simple shift from quantity to quality, but rather presents a dual challenge: it must ensure sustained food security in terms of quantity whilst simultaneously enhancing food safety standards. Taking this dual structure as its theoretical starting point, this study examines, in particular, the risk structure and institutional challenges associated with quality safety, as well as how these factors relate to the formation of consumer trust. Whilst this book focuses primarily on the "qualitative aspects" of food safety, a comprehensive understanding of food safety issues requires an integrated consideration of "quantitative aspects" such as the stability of the food supply and supply–demand structures.

The exploration of this issue will be conducted systematically in our forthcoming book, which is provisionally titled "*Food Security Issues in China: A Quantitative Perspective*". Specifically, the present study will re-examine the question of "who will feed China", presenting an analytical framework that takes into account changes in demographic trends and the structure of food demand. The study will also examine supply risks arising from agricultural productivity, resource constraints, climate change and the international environment from a medium- to long-term perspective. In addition, with reference to the roles of national strategy and technological innovation, the following discussion will clarify the realistic choices between complete self-sufficiency and strategic self-sufficiency, as well as the structure of a Chinese model of food security. On this basis, an attempt will be made to construct a comprehensive analytical framework for food safety issues by integrating the interrelationship between food safety (qualitative) and food security (quantitative) on theoretical grounds.

Consequently, whilst premising the importance of quantitative security, this book theoretically organizes the risk structure inherent in qualitative security and examines how this relates to the formation and erosion of consumer trust. The objective of this study is to reposition food safety issues so that they are recognized not only as technical or administrative challenges, but also as a matter of trust constructed within social interactions.

Food safety issues in China unfold within a dual framework of quantitative and qualitative safety; however, in modern society, ensuring qualitative safety in particular tends to converge on the issue of "trust". Specifically, this encompasses institutional trust, the belief that systems are functioning properly; organizational trust, the belief that companies are complying with regulations; and consumer trust, the belief that consumers can understand information and make rational choices.

The "three-layered trust model" presented in Chapter 2 of this book is a theoretical framework for reconstructing food safety, centered on the interaction between institutional trust, organizational trust and interpersonal/consumer trust. A key feature of this model is that it views the risk structure associated with qualitative safety as a dynamic process involving the generation, erosion and reconstruction of trust. In other words, risks in food safety should not be reduced merely to the hazards inherent in the food itself, but should be understood as a social phenomenon that is amplified or mitigated within a three-layered relationship of trust through factors such as information asymmetry, uncertainty in the operation of systems, and distrust of corporate behavior.

In the following section, drawing on the theoretical framework outlined above, this book will examine the current state of food safety issues in China and elucidate their structural characteristics by analyzing major incidents and shifts in public concern.

1.1.2 The Current State and Challenges of Food Safety in China

1.1.2.1 Major Food Safety Incidents in China and Changes in Public Concern

Table 1.1 summarizes major food safety incidents in China and their institutional implications, based on prior research and official sources (Pei et al., 2011; Gale & Hu, 2012). Furthermore, with regard to the institutional background and policy trends, Table 1.1 has drawn upon legal and regulatory documents published by the State Council of the People's Republic of China (2019) and the National People's Congress of China (2015). In addition, when examining specific cases, reports from international organizations such as the World Health Organization (2009).

As illustrated in Table 1.2, the results of a survey on the "Top 10 Most Notable Issues in China (2005–2022)" are summarized. This survey was conducted by publications such as the monthly magazine *Xiaokang*, which is published under the auspices of *Qiushi*, the official theoretical journal of the Communist Party of China (The Paper, 2022; Xiaokang Magazine, 2005, 2007). The survey was initiated in 2005 and comprised a total of 28 items, encompassing food safety, housing reform, environmental protection, energy and resource shortages, healthcare system reform,

Table 1.1 Major food safety incidents in China and their principal impacts and institutional consequences (1990s–2024)

Year(s)	Incident name	Description	Major impact/ Institutional outcome
1990s	Mass Methanol Poisoning from Counterfeit Liquor	Methanol mixed into counterfeit alcohol, causing deaths nationwide	Strengthening of food hygiene supervision; highlighting local administrative responsibility issues
2002	Pesticide Residue in Frozen Chinese Spinach	Detection of residual pesticides (toxic insecticides) in spinach exported to Japan	Strengthening of Japan's import regulations; review of China's export controls
2004	Fuyang Adulterated Milk Powder Incident	Infant deaths and developmental disabilities caused by adulterated and low-nutrition milk powder	Turning point in food safety policy; strengthening dairy product supervision
2005	Sudan Red Dye Incident	Use of industrial dyes in food (seasonings, processed foods)	Strengthening of additive regulations and enforcement against illegal additives
2005	Malachite Green (Aquatic Product) Incident	Use of banned drugs (malachite green) in aquaculture	Strengthening of aquatic product export regulations; inspection system improvement
2007	Melamine-contaminated Pet Food Incident	Melamine contamination in Chinese raw materials (e.g., gluten)	International food safety concern; decline in confidence in Chinese raw materials
2008	Sanlu Melamine Milk Powder Incident	Melamine-contaminated milk products causing approximately 300,000 victims	Trigger for national-level crisis; direct driver of food safety law reform
2008	Melamine-contaminated Eggs Incident	Melamine detected in eggs via feed	Strengthening of feed management and supply chain supervision
2008	Frozen Dumpling Pesticide Incident	Poisoning in Japan caused by organophosphate pesticides from Hebei frozen dumplings	Diplomatic issue between Japan and China; strengthening of export food traceability
2010	Gutter Oil Incident	Illegal circulation of recycled waste oil for food use	Expansion of social distrust; strengthening of criminal and circulation supervision

(continued)

Table 1.1 (continued)

Year(s)	Incident name	Description	Major impact/ Institutional outcome
2011	Clenbuterol Residue in Pork Incident	Residue of clenbuterol detected in pork from major companies	Collapse of trust in meat industry; strengthening corporate responsibility and inspection systems
2012	Chromium Gelatin Capsule Incident	Use of industrial chromium in pharmaceutical gelatin capsules	Strengthening of food and pharmaceutical boundary management
2013	Fake Mutton Incident	Sale of rat and fox meat as mutton	Strengthening of food fraud regulations; institutionalization of public security intervention
2013–2015	Cadmium-contaminated Rice Incident (Hunan Province)	Circulation of heavy metal contaminated rice (industrial pollution origin)	Highlighting necessity for integrated management of environment and food safety
2014	Shanghai Fuxi Expired Meat Incident	Supply of expired meat to food service chains	Trust crisis in foreign supply chains; strengthening supervision
2015	Food Safety Law Amendment	Revision of the food safety law, regarded as the strictest ever	Strengthening of penalties, risk management, and traceability systems
2016	Smuggled Powdered Milk Incident	Sale of counterfeit imported powdered milk (reported by EC)	Strengthening of cross-border regulations and brand certification systems
2019	School Cafeteria Safety Incidents (e.g., Chengdu)	Discovery of spoiled and unhygienic food	Strengthening of school meal supervision; local government accountability
2020	COVID-19 Detection on Frozen Food Packaging	Detection of virus on imported frozen food packaging	Strengthening of cold chain logistics management and inspection systems
2023	Jiangxi "Rat Head Duck Neck" Incident	Administrative explanation issue over foreign object contamination in school cafeteria	Risk communication challenges in the SNS (social media) era

Note This table is compiled by the author with reference to the aforementioned prior studies, government-published materials, reports from international organizations, and media sources employment issues, education costs, basic education in rural areas, social security, public safety, social credit, government public services, corruption, government system reform, regional development disparities, and wealth inequality. The survey was conducted online, and the evaluations were based on valid responses collected from over 10,000 participants. Despite the ongoing conduct of social and indicator surveys by *Xiaokang*, the specific survey format titled "Top 10 Most Notable Issues in China" has been discontinued since 2023.

According to the opinion survey results shown in Table 1.2, among the top 10 most notable issues in China, food safety ranked fifth in 2005, eighth in 2007, second in 2009, fourth in 2010 and third in 2011, before occupying the top spot for six consecutive years from 2012 to 2017. Although it subsequently dropped to fourth place in 2018 and fifth place in 2019, it rose to second place in 2020, regained the top spot in 2021, and stood at fourth place in 2022.

As can be seen from Table 1.2, public interest in food safety issues has shown a very distinctive trend. In this study, the key points discernible from the table into the following four categories are organized and they are analyzed from three perspectives: public concern, incidents and policy.

1.1.2.2 A Four-Stage Analysis of Food Safety Issues in China

Based on the major food safety incidents in China and the evolution of public concern outlined in Sect. 1.1.2.1, it can be understood that China's food safety issues have developed not merely as an accumulation of isolated incidents, but through the interplay of institutional changes, public perception and the information environment. In other words, they have evolved in stages through a cyclical process whereby the occurrence of food safety incidents leads to a decline in public trust, which in turn prompts policy responses and institutional reforms.

From this perspective, the following outlines the historical development of food safety issues in China, dividing it into four stages.

Phase 1: latent problem period (until 2008)

The period from the 1990s to 2008 can be characterized as a "latent problem period", during which food safety issues already existed but had not yet fully come to the fore socially or politically. During this period, food supply in China expanded rapidly alongside the progress of market-oriented reforms; however, the regulatory framework was multi-departmental and decentralized, with supervisory authority dispersed across multiple agencies such as the Ministry of Agriculture, the Ministry of Health and the General Administration of Quality Supervision, Inspection and Quarantine, and supervisory capacity was insufficient. This institutional structure led to inconsistencies in policy implementation, a blurring of responsibilities and delays in information sharing; as a result, responses to food safety issues remained largely reactive and symptomatic in nature.

During this period, incidents such as the counterfeit alcohol scandal and issues concerning pesticide residues occurred sporadically; however, these were treated as

Table 1.2 Top 10 most notable issues in China (2005–2022)

Rank	2005	2007	2009	2010	2011	2012	2013	2014	2015	2016	2017	2018	2019	2020	2021	2022
1	Health-care system reform	Health-care system reform	Health-care system reform	Prices (inflation)	Housing prices	*Food safety*	*Food safety*	*Food safety*	*Food safety*	*Food safety*	*Food safety*	Housing prices	Elderly care policy	COVID-19	*Food safety*	COVID-19
2	Environmental protection	Prices	*Food safety*	Housing prices	Prices	Prices	Pollution issues	Pollution issues	Health-care reform	Elderly care policy	Housing reform	Elderly care policy	Health-care reform	*Food safety*	COVID-19	Economic slow-down
3	Social security	Real estate prices	Corruption issues	Health-care reform	*Food safety*	Pollution issues	Health-care reform	Prices	Pollution issues	Housing prices	Environmental protection	Health-care reform	Prices	Social security	Education reform	Employment issues
4	Corruption	Income inequality	Environmental protection	*Food safety*	Health-care reform	Health-care reform	Income inequality	Housing prices	Income inequality	Healthcare reform	Housing prices	*Food safety*	Housing reform	Employment issues	Energy supply	*Food safety*
5	*Food safety*	Corruption	Social security	Education reform	Pollution issues	Housing reform	Health-care reform	Housing prices	Health-care reform	Pollution issues	Income distribution reform	Education reform	*Food safety*	COVID-19	Personal data protection	Education reform
6	Education costs	Environmental protection	Housing reform	Housing reform	Income inequality	Social security	Income inequality	Employment issues	Environmental protection	Environmental protection	Prices	Social trust	Public life & mental well-being	Personal data protection	Housing reform	Social trust
7	Social governance	Social security	Employment issues	Social security	Social ethics	Social security	Prices	Environmental protection	Prices	Disease control & public health	Social trust	Employment issues	Education reform	Education reform	Social trust	Personal data protection

(continued)

Table 1.2 (continued)

Rank	2005	2007	2009	2010	2011	2012	2013	2014	2015	2016	2017	2018	2019	2020	2021	2022
8	Real estate prices	*Food safety*	Education reform	Employment issues	Education reform	Education reform	Environmental protection	Employment issues	Social security	Prices	Health-care reform	Environmental protection	Income distribution reform	Health-care reform	Anti-monopoly	Health-care reform
9	Social morality	Employment	Social governance	Income distribution reform	Rising living costs	Income distribution reform	Social security	Environmental protection	Employment issues	Education reform	Pollution issues	Employment issues	Elderly care policy	Aging	Family education	Social security
10	Income inequality	Social governance	Democracy & rule of law	Pollution issues	Employment issues	Housing reform	Housing reform	Social ethics	Education reform	Social security	Elderly care policy	Prices	Housing prices	Public life & mental well-being	Family education	Aging

Source Xiaokang Magazine (2005, 2007)

isolated incidents and did not lead to fundamental reform of the system as a whole. In other words, systematic risk assessment was underdeveloped and institutional integration was inadequate. In particular, the prioritization of economic growth by local governments, a lack of regulatory resources, and information asymmetry combined to cause problems such as the use of illegal additives and quality misrepresentation to accumulate structurally. This situation could be described as a "low-trust, low-regulation equilibrium"; although food safety risks existed, they had not yet been recognized as a systemic problem by society as a whole (Gale & Hu, 2012; Lam et al., 2013).

Nevertheless, social anxiety was gradually building even at this phase. For example, food safety issues ranked fifth in 2005 and eighth in 2007 in social issue rankings, confirming that they were attracting a certain level of attention. On the other hand, when compared to lifestyle-related issues such as healthcare, prices and housing, its priority was relatively low, and food safety was perceived as a "moderate concern". Nevertheless, it can be said that food safety was already beginning to be gradually recognized as one of society's key issues.

Phase 2: crisis escalation period (2008–2012)

The Sanlu infant formula scandal (melamine contamination incident) that occurred in China in 2008 marked a major turning point that transformed China's food safety issues into a social crisis. In this incident, the industrial chemical melamine was illegally added to infant formula manufactured by the Sanlu Group in Hebei Province, causing health problems for numerous infants and resulting in reported deaths (World Health Organization, 2009; Pei et al., 2011). Melamine was used to conceal the falsification of protein content through the dilution of raw milk, exposing deficiencies in the supervisory system and a lack of corporate ethics throughout the supply chain (Pei et al., 2011). Furthermore, this incident marked a turning point in food safety governance in China, prompting the government to amend food safety laws and strengthen the supervisory system. On the other hand, institutional trust among consumers was severely undermined, and it had a long-term impact on the structure of the domestic dairy market and consumer behavior (Lam et al., 2013).

This incident made it clear that food safety issues were not merely a matter of misconduct by individual companies, but rather a national risk stemming from systemic flaws (such as fragmented oversight, local protectionism and the concealment of information); it marked a critical juncture in China's food safety governance (Pei et al., 2011).

In response to this crisis, the Chinese government enacted the Food Safety Law in 2009 and initiated institutional reforms, including the establishment of the State Council Food Safety Commission and the introduction of risk assessment and risk management systems; however, in the early stages, these measures failed to yield sufficient results due to a fragmented supervisory framework and a lack of enforcement capacity (Lam et al., 2013; Liu et al., 2019). At the same time, fraudulent practices continued to occur even after the melamine scandal, and consumer distrust was, if anything, further reinforced (Qiao et al., 2012).

This marked a shift from the traditional "reactive" approach to a risk-management-based governance model grounded in the precautionary principle. Furthermore, the whole-food-chain management approach advocated by the FAO and WHO began to be adopted internationally (FAO & WHO, 2003). Therefore, this period can be described as a transitional phase in which risks not only "materialized" but were also recognized as a "social crisis", prompting the initiation of institutional reforms.

Phase 3: peak of interest period (2012–2017)

Between 2012 and 2017, food safety issues were consistently recognized as the most pressing concern in Chinese society, ranking as the number one topic of public interest for six consecutive years. This period marked a stage in which food safety was perceived not merely as a series of isolated incidents, but as a "structural risk" that undermined the trust of society as a whole, leading to the emergence of a crisis of social trust. This sustained high level of concern can be explained not merely by the frequent occurrence of incidents, but by the combined effects of institutional, social and psychological factors.

Firstly, the 2008 melamine-tainted infant formula scandal affected more than 300,000 infants and dealt a severe psychological blow to Chinese society. This incident exposed not only corporate malpractice but also flaws in the regulatory system and delays in government information disclosure, thereby fundamentally exacerbating institutional distrust. Subsequently, a series of scandals, including the "gutter oil" scandal and food mislabeling, led to food safety issues being recognized not as exceptional risks but as structural problems (Klein, 2013; Lam et al., 2013; Liu et al., 2019).

Secondly, constraints in institutional governance contributed to the protracted nature of the problem. At the time, food safety management was fragmented across multiple agencies, and this fragmentation of oversight undermined the effectiveness of regulation. Combined with local governments' prioritization of economic growth and a lack of supervisory resources, the deterrent effect against violations was limited, and an incentive structure characterized by "low costs and high profits" was maintained (Gale & Hu, 2012; Lam et al., 2013). In response, the Chinese government pursued institutional reforms and regulatory tightening, including the enactment of the Food Safety Law in 2009, the establishment of the China Food and Drug Administration (CFDA) in 2013, and the implementation of the amended Food Safety Law in 2015. However, a paradoxical phenomenon arose: as the number of crackdowns and media reports increased during the process of institutional improvement, the public perception that "the problem is getting worse" was actually reinforced (Kasperson et al., 1988; Lam et al., 2013).

Thirdly, urbanization and the increasing complexity of supply chains have heightened risk perception. As food distribution has expanded across wider areas and reliance on processing has grown, supply chains have become longer, highlighting issues such as a lack of traceability and difficulties in management. These structural changes have not only altered the reality of food risks but have also increased consumers' perception of uncertainty (Ortega et al., 2011).

Fourthly, the explosion of social media has led to the social amplification of risk perception. With the expansion of platforms such as Weibo and WeChat, information regarding food safety has spread rapidly; the sharing of such information, including unverified details and rumors, has tended to amplify risk perception beyond actual levels. This is a typical phenomenon explained by the "theory of social amplification of risk" (Kasperson et al., 1988). Furthermore, delays and a lack of transparency in the government's response accelerated the decline in trust in the system through this amplification process.

Fifthly, the expansion of the middle class and shifts in consumer attitudes played a significant role. As income levels rose, Chinese consumers increasingly prioritized safety and quality over price, and food safety evolved from a mere hygiene issue into a matter concerning "quality of life" and "social trust" (Liu et al., 2013; Ortega et al., 2011).

Taking all of the above into account, the reason why food safety issues remained at the forefront of public concern from 2012 to 2017 was not merely due to the frequent occurrence of food safety incidents, but rather the result of the interplay between institutional distrust, changes in the supply structure, the evolution of the information environment, and a shift in consumer attitudes. This period can be described as a process in which "objective increases in risk" and "subjective breakdowns in trust" occurred simultaneously, representing the manifestation of a classic systemic risk (a risk in which individual risk events interlink and amplify one another, leading to the dysfunction or collapse of the entire system). Furthermore, food safety issues came to be positioned, alongside other social challenges such as corruption, environmental pollution and healthcare reform, as matters closely linked to the quality of governance and institutional trust.

Phase 4: policy response and scattered attention period (2017–present)

As shown in Table 1.1, following the comprehensive revision of the Food Safety Law in 2015 and the subsequent integration of regulatory systems, China's food safety governance has transitioned into a "mature phase". This revision introduced principles such as "prevention first", "risk management" and "full-process supervision", bringing the regulatory framework closer to international standards. Furthermore, the supervisory authority was reorganized from the CFDA (China Food and Drug Administration) to the SAMR (State Administration for Market Regulation). Combined with the introduction of traceability systems and digital supervision, this marked a shift from the previous decentralized management system to an integrated, risk-based management system (National People's Congress of China, 2015; State Council of the People's Republic of China, 2019).

Whilst these institutional measures have made it possible to manage food safety risks to a certain extent, problems such as the misuse of food additives and pesticide residues remain. On the other hand, changes can be observed in terms of public interest. Since 2018, interest in food safety issues has declined relatively, and a "dispersion of interest" has progressed due to competition with other social issues such as environmental problems, healthcare reform, housing issues, employment, and even the novel coronavirus disease (COVID-19) (The Paper, 2022).

There are three main factors behind this change.

Firstly, the government has progressively tightened institutional regulations. Since the enactment of the Food Safety Law in 2009, measures such as the establishment of a supervisory framework, the introduction of a food traceability system, and the strengthening of real-time monitoring systems utilizing the internet have been implemented in stages. Furthermore, the establishment of the SAMR in 2018 centralized food safety supervision functions, a move widely regarded as having enhanced the reliability of the institutional management framework (Liu et al., 2019; The State Council of the People's Republic of China, 2019).

Secondly, society's priority issues themselves have become more diverse. Measures to address an ageing population (elderly care policies), healthcare reform, soaring house prices, employment issues, and the emergence of the COVID-19 pandemic from 2020 onwards have all come to the fore, dispersing public attention. In particular, COVID-19 has reshaped public concern regarding health risks in general, leading to food safety issues and infectious disease control being perceived as closely interlinked. Consequently, it is reported that since 2018, public satisfaction and interest in food safety have fallen to around fourth or fifth place in the overall rankings (The Paper, 2022).

Thirdly, the nature of food safety issues themselves has changed. Although the frequency of major incidents, such as those seen between 2011 and 2017, has decreased, these issues are increasingly being viewed as 'chronic social challenges' requiring ongoing attention, covering areas such as the management of the entire food supply chain, quality data management utilizing digital technology, and the nature of corporate self-regulation. Some food companies and regional initiatives are making progress in introducing traceability systems using blockchain and IoT, demonstrating the potential for new supervisory methods such as the visualization of quality information and real-time monitoring. Although these are not central to institutional reform, they are attracting attention as complementary tools for preventive supervision (FAO & WHO, 2003; The State Council of the People's Republic of China, 2019).

In light of the above, China's food safety issues can be categorized into the following four stages of structural development: (1) latent problem period (where risks exist but are invisible); (2) crisis escalation period (where institutional flaws were exposed by the melamine scandal); (3) peak of interest period (where social amplification has cemented the issue at the top of the public agenda); and (4) policy response and scattered attention stage (where risks remain, albeit manageable, due to institutional reforms). This four-stage model is consistent with the theory of social amplification of risk (Kasperson et al., 1988) and theories of institutional change, providing a framework for understanding China's food safety issues not merely as a succession of incidents, but as a "dynamics of trust and institutions".

Consequently, food safety is no longer merely a matter of consumer protection or hygiene management, but is positioned as a comprehensive indicator reflecting the state's governance capacity, industrial structure, and the level of social trust. This perspective offers important insights for analyzing future institutional reforms and innovation policies in China.

1.1.3 An Analysis of Food Safety Issues in China from the Perspectives of Public Concern, Incidents and Policy

Food safety issues in China are not merely a collection of isolated incidents, but a complex phenomenon that reflects the overall structure of trust within society and institutional changes. Firstly, repeated food safety incidents have not only caused harm to public health but have also led to a loss of trust in corporate ethics and government oversight and have come to be recognized as a social risk. Secondly, looking at trends in public concern, food safety has been positioned as the most pressing social issue for a certain period, and social concern has intensified alongside growing risk awareness. Furthermore, this accumulation of public concern and incidents has served as a strong catalyst for policy responses, such as the enactment and amendment of food safety laws and the reorganization of supervisory bodies.

Therefore, to understand China's food safety issues, it is necessary to take an integrated view of three aspects: (1) the occurrence of incidents and their structure; (2) changes in risk perception and trust driven by public concern; and (3) policy and institutional responses. It is precisely the interaction between these three elements that has constituted the crucial backdrop defining the evolution of food safety governance in China.

1. The public perspective: A shift from a single-issue focus to a multifaceted understanding of risk

Changes in public concern in China should be viewed not merely as a shift in focus, but as a "structural change in risk perception". As shown in Tables 1.2 and 1.3, in the late 2000s, the primary areas of concern were "livelihood-related risks" directly linked to the cost of living, such as rising prices, the healthcare system and housing prices. These concerns arose against a backdrop of increasing living costs driven by rapid economic growth and urbanization and were recognized as issues directly affecting individuals' livelihood stability. At this stage, whilst food safety attracted a certain degree of attention, it remained a relatively low priority and was regarded merely as a "potential risk".

However, by the early 2010s, food safety issues had rapidly come to the forefront of public concern. The essence of this shift lies in the transition from "individual risk" to "trust risk". In other words, food safety came to be recognized not merely as a matter of health hazards, but as a question of social trust in institutions and the market. Following the 2008 melamine-tainted infant formula scandal, a succession of food safety incidents led to a rapid increase in consumer distrust, and behavioral changes such as a preference for imported foods and an increase in cross-border shopping have been observed (Pei et al., 2011).

Furthermore, from 2018 onwards, public interest underwent another significant structural shift. In addition to an ageing population (pension issues), healthcare reform and housing problems, new challenges such as the COVID-19 pandemic and personal data protection emerged simultaneously, causing social concerns to shift

Table 1.3 Analysis from the perspectives of public concern, incidents and policy

Stage	Period	Characteristics of public concern	Representative incidents & social issues	Policy responses & institutional Changes
Latent Problem Stage	1990s–2008	Public concern focused on economic growth, prices, and employment. Limited attention to food safety; risk perception was low and fragmented	Fake alcohol (methanol) poisoning, pesticide residues, illegal use of additives occurred sporadically	Strengthening of food hygiene management began, but regulatory authority remained fragmented across multiple agencies; integrated systems were not yet established
Crisis Outbreak Stage	2008–2012	Public concern centered on prices, healthcare, and housing; food safety issues gradually emerged as a major concern	Melamine-contaminated milk scandal (2008) led to a sharp increase in public distrust of food safety	Enactment of the Food Safety Law (2009) and initiation of regulatory reforms
High Concern Stage	2012–2017	Food safety became the most prominent social issue; consumer anxiety intensified	"Gutter oil" scandal, toxic bean sprout incidents, "lean meat powder" (clenbuterol) scandals, among others	Establishment of the China Food and Drug Administration (CFDA) (2013); revision of the Food Safety Law (2015)
Policy Response & Diversification of Public Attention Stage	2018–Present	Declining concern over food safety; rising attention to aging, healthcare reform, and COVID-19	COVID-19 pandemic, energy issues, personal data protection concerns	Establishment of the State Administration for Market Regulation (SAMR) (2018); institutional integration and strengthening of regulatory systems

Note This table was compiled and prepared by the author based on official statistical data, relevant policy documents (from the National People's Congress, the State Administration for Market Regulation, the State Council, etc.), and existing research regarding the development of food safety issues in China. The time periods and characteristics of each stage are categorized for analytical convenience from a "single-issue focus" to a "multi-issue, dispersed focus" (The Paper, 2022). At this stage, whilst food safety remains a significant issue, it is now positioned on a par with other systemic risks. This suggests that Chinese society has transitioned into a more complex and multi-layered risk environment.

2. The incident perspective: from isolated accidents to the visualization of structural risks

Perceptions of food safety issues in China have shifted from "isolated incidents" to "structural risks" as a series of incidents have accumulated. A particular turning point was the 2008 Sanlu infant formula scandal. In this incident, the toxic substance melamine was mixed into infant formula, causing health problems for a large number of infants. This incident went beyond mere corporate malpractice; it exposed the fragility of the entire supply chain and the inadequacies of the regulatory system, serving as a catalyst for society to strongly recognize food safety issues as a systemic challenge (Pei et al., 2011).

Subsequently, as the 2010s began, a series of food safety incidents occurred, including the "gutter oil" scandal, cases involving toxic foodstuffs, and the counterfeit meat scandal. A defining feature of these incidents was that they were not confined to specific companies or products, but came to be recognized as risks affecting the food industry as a whole. In other words, the incidents occurred in a chain reaction characterized by "recurrence" and "diversity", cumulatively amplifying consumer distrust. As a result, food safety emerged as one of the most critical public issues in Chinese society.

Furthermore, the spread of the novel coronavirus disease (COVID-19) around 2020 expanded the very nature of the risks involved. Whereas traditional food safety issues were primarily risks stemming from industry and distribution, the pandemic, as a biological and global risk, highlighted the need for crisis management at the national level. In this process, the scope of food safety expanded beyond mere hygiene management to encompass areas closely linked to "public health" and "national security".

As outlined above, it can be said that food safety issues in China have developed within a dynamic in which the accumulation of individual incidents has fueled growing public mistrust, which in turn has prompted institutional reforms and a shift in risk perception.

3. The policy perspective: from a fragmented approach to integrated governance

Food safety policy in China has evolved gradually in response to public pressure and a series of incidents. Its essence is understood not merely as institutional reform, but as a "reconstruction of governance capacity".

In the initial phase, as evidenced by the enactment of the Food Safety Law in 2009 shown in Table 1.1, institutional development was limited to piecemeal and reactive measures. This period was characterized by a decentralized supervisory system involving multiple agencies, including the Ministry of Agriculture, the General Administration of Quality Supervision, Inspection and Quarantine, the State Administration for Industry and Commerce, and the Ministry of Health, with structural problems such as overlapping jurisdictions and ambiguous responsibilities becoming apparent.

However, in response to the frequent occurrence of food safety incidents and growing public concern during the 2010s, there was a significant shift in policy. The establishment of the China Food and Drug Administration (CFDA) in 2013 and the implementation of the amended Food Safety Law in 2015 symbolized the strengthening of regulations and the reorganization of the supervisory framework. At this stage, the effectiveness of the system was significantly enhanced through measures such as stricter penalties, the clarification of corporate responsibility, and the introduction of risk management (Pei et al., 2011; Liu et al., 2019).

Furthermore, since 2018, policy has entered a phase of "integration and advancement". With the establishment of the SAMR, market supervision functions, including food safety, were centralized, and the traditional siloed structure was significantly overhauled. This reform signified a shift from decentralized governance to integrated governance and is regarded as having contributed to improved efficiency and consistency in the operation of the system (World Bank, 2021).

At the same time, policy instruments are becoming more sophisticated. Through the use of big data and the development of traceability systems, supervision is shifting from the traditional reactive approach to a preventive and continuous management system that prioritizes the early identification of risks. However, these digital technologies are positioned as complementary tools rather than the core of the system, and the strengthening of governance capacity continues to rely on legal frameworks and organizational restructuring.

As outlined above, it can be said that China's food safety policy has evolved from a "decentralized response" to "integrated governance", aiming to alleviate social distrust and restore confidence in the system. At the same time, efforts are being made to address emerging risks such as an ageing population and pandemics, and food safety policy is being repositioned as part of broader public governance.

Taking all of the above into account, it can be concluded that food safety issues in China have developed through the following dynamic mechanism:

1. Incidents make risks visible;
2. Public opinion amplifies them as social problems;
3. Policies respond institutionally.

More importantly, this process is not unidirectional but unfolds in a "cyclical and evolutionary" manner. Whilst policy tightening may temporarily reduce public interest, the emergence of new incidents or risks reignites public concern, prompting further institutional reform.

1.2 An Analysis of Factors Contributing to the Growing Public Concern for Food Safety

Below, the factors behind the growing social concern for food safety are analyzed from four perspectives.

1.2.1 Rising Income Levels and the Expansion of the Middle Class

Rising income levels and the expansion of the middle class in China have significantly altered the nation's consumption patterns and attitudes towards food. As incomes rise, the Engel coefficient, which indicates the proportion of household expenditure spent on food, tends to fall; this is used as an indicator reflecting improvements in quality of life and the diversification of consumption. In China, according to reports based on data from the National Bureau of Statistics, the Engel coefficient fell from 30.1% to 28.2% during the 13th Five-Year Plan period (2016–2019), indicating that, alongside the sustained increase in personal disposable income, the proportion of expenditure on items other than food has risen. This decline demonstrates the improvement in living standards and the sophistication of consumption patterns.

At the same time, the expansion of the middle-income group is a key factor driving increased interest in quality safety within Chinese society. Analyses based on data from the National Bureau of Statistics and various surveys indicate that the proportion of the middle-income group in China's urban areas amounts to approximately 66.5% of the total population defined by disposable income, and it is estimated that the middle class accounts for around 30 to 40% of the total urban population. This reflects a shift away from the traditional social structure centered on low-income groups towards a broader base of income levels and is having a significant impact on people's consumption patterns and values (Kharas, 2017; National Bureau of Statistics of China, 2023).

The expansion of this middle-income group is accompanied by social implications that go beyond mere increases in income. As incomes rise, consumers tend to shift from price-conscious, basic subsistence consumption towards "value-added" consumption focused on quality, safety and health. This trend is evident in the growing demand for health foods, organic foods and high-value-added products as income levels rise, as demonstrated by the differences in consumption patterns among the middle class in the daily necessities market. Furthermore, this shift towards more sophisticated consumption is directly linked to concerns about food safety, acting as a driving force behind heightened societal demands for food safety and quality control.

Rising income levels and the expansion of the middle class are driving a shift in consumer consciousness from a focus on securing sufficient food quantities to a focus on food safety and health, and this situation is influencing the direction of food safety policies and regulations. Whilst quantitative issues remain of fundamental importance, the growing demand for "safe and reliable food" driven by the rise of the middle class is prompting a renewed institutional recognition of the importance of food safety and quality in China.

1.2.2 A Stronger Focus on Health

The growing social concern for food safety in China is manifesting itself in the form of a heightened focus on health. As the economy grows and income levels rise, consumer priorities are shifting from "quantitative sufficiency" to "leading a healthy lifestyle". This shift is closely linked to changes in dietary patterns and the rise in chronic diseases.

According to a report by the World Health Organization (WHO), non-communicable diseases (NCDs) account for approximately 90% of all deaths in China, with cardiovascular disease, diabetes and cancer being the leading causes of death (World Health Organization, 2022). It has been suggested that this shift in the disease profile is linked to diets becoming higher in fat and sugar, and an increasing reliance on processed foods. Indeed, statistics from the FAO and the OECD show that per capita meat consumption in China has surged since the launch of the reform and opening-up policy, reaching a level exceeding 60 kg per year in the 2020s (OECD & FAO, 2023). Whilst this expansion in the consumption of animal-based foods has contributed to improved nutrition, it is also linked to an increased risk of obesity and lifestyle-related diseases.

The institutionalization of health-conscious policies is also progressing at the national level. The Healthy China 2030 plan, published by the Chinese government in 2016, positions the improvement of public health as a national strategy, setting nutritional improvement, enhanced food safety and the prevention of chronic diseases as key objectives (State Council of the People's Republic of China, 2016). This policy redefines food safety not merely as a matter of hygiene management, but as an integral part of public health policy and sustainable development strategy.

Furthermore, according to the National Bureau of Statistics of China, per capita healthcare expenditure has been on a long-term upward trend, and the proportion of healthcare-related costs within household expenditure is also rising (National Bureau of Statistics of China, 2023). This serves as an empirical indicator of growing concern for health at the household level. Market data also confirms growing demand for health foods, organic foods, and low-sugar and low-fat products, indicating a heightened focus not only on food safety but also on "functionality" and "nutritional value".

Thus, the strengthening of health-consciousness in Chinese society is underpinned by multiple factors, including rising incomes, changes in the disease profile, and shifts in national strategy. Food safety is no longer confined to accident prevention and regulatory compliance; it has expanded into a comprehensive concept linked to health promotion, disease prevention and the improvement of quality of life. Consequently, ensuring food safety should be positioned as an integrated challenge that bridges food governance and public health policy.

1.2.3 The Social Impact of Food Incidents

When discussing the growing public concern regarding food safety in China, the social impact of food safety incidents has been immense. In particular, the 2008 Sanlu infant formula scandal sent shockwaves through Chinese society and served as a catalyst for redefining food safety as a national priority. In this incident, the contamination of infant formula with melamine caused health problems for a large number of infants; according to official reports, approximately 300,000 infants and young children were affected (World Health Organization, 2009). This incident went beyond a mere corporate scandal, fueling widespread distrust in the entire system of governance, including regulatory frameworks, the accountability of local governments, and the handling of information disclosure.

Since then, reports of food safety incidents and risks have continued to attract widespread public attention. For example, the 2011 "gutter oil" scandal (involving the reuse of waste cooking oil), as well as cases of pesticide residues and heavy metal contamination, have highlighted the risks lurking at every stage of the supply chain. These issues have made the public realize that food safety must be managed throughout the entire process, not only at the production stage, but also during processing, distribution and sale.

Internationally, concerns regarding food safety in China have been repeatedly raised as well. The World Bank has pointed out that strengthening food safety governance in China is essential for restoring consumer confidence and ensuring market sustainability (World Bank, 2021). Indeed, following food safety incidents, mistrust towards specific brands and domestic products has grown, leading to noticeable changes in consumer behavior, such as increased demand for imported goods and the expansion of cross-border e-commerce.

In terms of institutional frameworks, the Food Safety Law came into force in 2009 and underwent major revisions in 2015 and 2018. Furthermore, the establishment of the SAMR in 2018 led to the reorganization of previously fragmented supervisory functions. These institutional reforms can be understood as policy measures aimed at restoring public confidence in the wake of food safety incidents.

Between October and November 2022, *Xiaokang*, in collaboration with the National Information Centre, conducted a survey on the "China Contemporary Comprehensive Development Index", which comprises 12 categories relating to the lives of Chinese citizens (food safety, consumption, public services, healthcare and health, ecological environment, social security, education, credit, housing, leisure, elderly care and sense of well-being) (The Paper, 2022).

When the results were ranked in order of satisfaction, public safety (law and order) came first (57.9%), followed by the ecological environment (55.0%) and consumer affairs (47.2%). Meanwhile, food safety ranked only seventh, with a relatively low satisfaction rate of 17.3%.

The specific rankings and satisfaction levels are as follows:

- Public safety (law and order) 57.9%
- Ecological environment 55.0%
- Consumer affairs 47.2%
- Healthcare and wellbeing 41.6%
- Residence 32.6%
- Education 20.5%
- Food safety 17.3%
- Leisure 15.9%
- Elderly care (welfare for the elderly) 8.7%
- Reputation 5.2%
- Public services 2.4%
- Sense of happiness 2.2%

1.2.4 Risk Amplification Through the Mass Media and Social Media

The growing public concern regarding food safety in China is closely linked to the development of mass media and social media. Since the late 2000s, the rapid increase in internet users and the widespread adoption of social media have enabled information on food safety to spread more quickly and extensively than ever before. According to a report by the China Internet Network Information Centre (CNNIC), the number of internet users in China exceeded 1 billion in 2023, with a penetration rate of approximately 75% (China Internet Network Information Center, 2023). This vast online space has become a primary forum for sharing and discussing information on food incidents and health risks.

In particular, social media platforms such as Weibo and WeChat have the capacity to instantly spread reports of food safety incidents and misconduct across the country. Since the 2008 Sanlu infant formula scandal, many food safety issues have come to light through online public debate and whistleblowing, prompting the authorities to take action. Through the interaction between media coverage and social media, risks tend to rapidly escalate from mere isolated incidents into social issues.

Research on risk communication has pointed out that risk perception tends to be "amplified" in digital environments. According to Kasperson et al.'s (1988) model of social risk amplification, the media serves not only as a channel for transmitting information but also as a mechanism for reinforcing and reconstructing the meaning of risk. In the case of China, food safety issues often become linked to emotional reactions and moral criticism on social media, which can lead to expressions of distrust in the system.

On the other hand, social media is also being utilized as a means for government agencies to promptly disclose information and conduct fact-checking. The SAMR and local governments publish inspection results and information on penalties through their official accounts in an effort to improve transparency. However, the oversupply of information and the spread of unverified information also carry the inherent risk of causing reputational damage and market disruption.

Thus, the media and social media simultaneously serve to both "visualize" and "amplify" food safety risks. Ensuring food safety in the digital age is closely linked not only to technical oversight systems but also to appropriate risk communication strategies, the reliability of information disclosure, and the maintenance of public trust. Therefore, food safety issues must be understood as social structural problems that are inseparable from the information environment.

1.3 The Risk Structure Related to Food Safety: A Multifaceted Analysis of Production, Processing, Distribution, and Oversight Systems

Food safety is not determined by a single factor but is shaped by the structural conditions of the entire supply chain, spanning production, processing, distribution, and consumption. In China in particular, the combined effects of rapid industrialization, urbanization, the expansion of distribution networks, and the advancement of the digital economy have made food safety risks more sophisticated and complex. In this section, we will categorize the major risks affecting food safety by stage and examine their social and institutional contexts.

1.3.1 Production Stage

The risk structure related to the quality and safety of food first becomes apparent at the production stage. In China, rapid agricultural intensification and the expansion of chemical inputs have led to structural risks such as pesticide residues, heavy metal pollution, and environmental degradation.

First, there is the problem of pesticide residues. Since the reform and opening up, the amount of pesticides used has increased with the aim of improving the yield per unit area. According to FAO statistics, China is one of the world's largest pesticide users and has a high level of use intensity per arable land (Food and Agriculture Organization of the United Nations, 2022). While pesticides contribute to production stabilization, improper use and residues that exceed norms can increase health risks for consumers. The Chinese government is developing residue standards (MRLs) and strengthening the inspection system, but the decentralized structure of smallholder farmers contains difficulties in management.

Second, there is the problem of heavy metal pollution. With the increasing industrialization and urbanization, the accumulation of cadmium, lead, mercury, etc. in the soil has become a problem. According to the National Soil Contamination Survey published by the Ministry of Environmental Protection and the Ministry of Land and Resources of China, about 16% of the survey sites were found to be contaminated beyond the standard (Ministry of Environmental Protection of the People's Republic of China & Ministry of Land Resources of the People's Republic of China, 2014). In particular, cadmium is highly migrated to rice and is directly related to the safety of staple foods, so it is of great social interest.

Third, there is the problem of environmental degradation. Excessive use of chemical fertilizers and pesticides, overuse of groundwater, and improper disposal of livestock waste have led to a decrease in soil fertility and deterioration in water quality. Environmental externalities in Chinese agriculture pose significant sustainability challenges. Environmental degradation is not limited to a single-year food safety problem, but is also a risk of eroding the food production base itself in the long term.

As mentioned above, quality safety risks in the production stage are not problems that can be reduced to the actions of individual farmers, but are deeply linked to institutional factors such as industrialized agricultural models, the effectiveness of environmental regulations, and local economic structure. Therefore, in order to ensure the quality and safety of food, it is essential not only to strengthen residue standards and expand inspections, but also to take structural measures such as switching to environmentally friendly agriculture and designing green production incentives.

1.3.2 Processing Stage

Risks related to the quality and safety of food are multi-layered not only at the production stage but also at the processing and distribution stages. In particular, microbial risks, inadequate hygiene systems, and inadequate cold chains are important structural factors that directly affect consumer health.

First, there is the issue of microbial risk. Globally, the main causes of food poisoning are pathogenic microorganisms such as Salmonella, Campylobacter, and Listeria. The World Health Organization estimates that approximately 600 million people worldwide suffer from diseases caused by contaminated food each year (World Health Organization, 2015). In China, bacterial food poisoning accounts for the majority of reported cases in the data published by the National Health Commission, and poor management in the group meal and restaurant industries is pointed out as a background factor. Poor temperature control and cross-contamination during the processing phase are major contributors to risk amplification.

Second, there is the issue of hygiene management. The size disparity in food processing enterprises is still large, and small businesses may not have sufficient implementation of management systems based on international standards, such as HACCP. Although China clarified the main responsibility of enterprises in the 2015 revision of the Food Safety Law and promoted the introduction of a traceability system, regional imbalances in supervision resources and transparency of information disclosure are still challenges (National People's Congress of the People's Republic of China, 2015). Although the system is being developed, ensuring effectiveness is the core of quality and safety.

1.3.3 Distribution and Sales Stages

In the distribution and sales stages, microbial contamination and poor hygiene management are important risks. Proper temperature control is essential, especially for fresh and ready-to-eat foods, and the lack of a cold chain increases the probability of food poisoning. With urbanization and changing consumption patterns, the risk of cross-contamination in the cooking and delivery process is gaining social attention as the use of eating out and delivery increases.

In recent years, the online food delivery market in China has been expanding rapidly. Meituan Inc., one of China's largest online food delivery platform companies, and Ele.me and others have greatly improved convenience, they have created new supervisory issues, such as the situation where consumers order and consume food through online platforms without directly confirming the actual cooking environment, hygiene conditions, and operators, or the problem of "invisible kitchens" such as cooking facilities and business formats, and the existence of unauthorized business outlets. While the digital economy enables information visualization, it also complicates the location of responsibility. More details will be discussed later in Chap. 7.

In addition, refrigerated and frozen logistics are essential for ensuring the safety of fresh and frozen foods, but regional disparities are noticeable in China. According to a report by the China Federation of Logistics and Purchasing (CFLP), although cold chain logistics has expanded rapidly in recent years, there are still challenges in rural and inland areas due to lack of equipment and transportation instability. If proper temperature control is not maintained, microbial growth and quality degradation can progress, making it a breeding ground for food accidents.

Especially during the COVID-19 pandemic, the issue of virus detection surrounding imported frozen foods has attracted social attention, reaffirming the fragility of cold chain management and the need to strengthen the supervision system. This indicates that the distribution stage is not just a logistics process but a critical risk management phase.

As mentioned above, the risk structure at the distribution and sales stages is formed by a complex combination of microbiological risks, institutional management capabilities, and the level of logistics infrastructure development. In order to ensure quality and safety, it is essential to have a "farm-to-table" perspective from production to consumption, and it is necessary to establish an integrated governance system (Food and Agriculture Organization of the United Nations & World Health Organization, 2003).

1.3.4 Supervision and Institutional Stages

Risks related to the quality and safety of food are highly dependent on structural factors not only at the production and distribution stages, but also at the supervision and institutional stages. In particular, regional differences in regulatory enforcement, local government incentive structures, and information disclosure challenges are important in understanding China's food safety governance.

First, there are regional differences in regulatory enforcement. In China, the Food Safety Law, which came into effect in 2009 and was amended in 2015, clarified the main responsibility of enterprises and the risk monitoring system, and the State Administration for Market Regulation (SAMR) was established through administrative reform in 2018. However, the actual supervision and enforcement is highly dependent on local governments. It has been pointed out that differences in financial strength and personnel structure can lead to regional differences in testing frequency and technical capabilities (The World Bank, 2021).

Second is the incentive structure of local governments. Under a decentralized economic system, local governments have placed emphasis on GDP growth and investment attraction. This growth-oriented incentive can in some cases lead to deregulation and oversight of local companies. The 2008 Sanlu infant formula scandal exposed not only corporate corruption but also local government supervisory responsibilities and information processing problems. In China's decentralized governance structure, there is an institutional tension between the policy goals of the senior government and the actual actions of local governments, and the field of food safety is no exception.

Third, there is the issue of information disclosure. As mentioned above, food products are goods that are difficult for consumers to fully evaluate and verify their quality, safety, and attributes not only before purchase but also after purchase, and have no choice but to rely on information and institutional guarantees from experts and suppliers, so-called typical "credence goods" (Darby & Karni, 1973), since consumers cannot check the quality in advance, transparent information disclosure by the government is the foundation for building trust. The 2015 revision of the Food Safety Law stipulates the strengthening of the recall system for the publication of illegal company names, but there is room for improvement in the speed of information disclosure and the degree of standardization (National People's Congress of China, 2015; State Council of the People's Republic of China, 2019).

As mentioned above, the risk structure at the supervision and institutional stages is defined not only by the development of the legal system, but also by regional disparities in enforcement capacity, evaluation incentives of local governments, and transparency of information disclosure. Ensuring quality security is an issue related to the credibility of governance and should be understood in the context of the interaction between institutional reform and trust-building.

1.4 Consumer Anxiety and Behavior Change: The Emergence of Trust Issues

1.4.1 Consumer Risk Perception

The issue of food quality and safety is defined not only by the presence or absence of objective danger, but also by the subjective aspect of how consumers perceive it. In other words, food safety issues expand their social impact not only through "risk itself" but also through "risk recognition". In China, since the Sanlu infant formula scandal in 2008, distrust of food safety has increased significantly, and consumers' risk perception structure has changed significantly.

According to the classical framework of risk perception research, people judge risk by different criteria than professional probability assessment. Food safety issues are characterized by direct health impacts, high frequency of daily contact, and information asymmetry, making them an area where cognitive risks are easily amplified. In addition, Kasperson et al.'s (1988) Social Risk Amplification Model describes the process by which media and interpersonal communication expand risk perception and amplify social impact. Information dissemination in China's social media environment is consistent with this theoretical framework.

Empirical studies have also shown that Chinese consumers' anxiety about food safety is at a high level. In particular, anxiety about food safety is easily associated with distrust of domestic foods, preference for imported foods, and reliance on prices and certification labels (Liu et al., 2013; Ortega et al., 2011). It has also been pointed out that food safety insecurity has had a long-term impact on the domestic dairy market and consumer behavior since the melamine milk powder incident (Pei et al., 2011; Qiao et al., 2012).

In addition, risk perception also varies according to socioeconomic attributes. Higher levels of education and income tend to have greater access to information and are more sensitive to risk, while lack of information can lead to underestimation in rural areas. Thus, risk perception is not uniform and involves differences embedded in the social structure. This part is further described in Chap. 3.

In general, the quality and safety of food in China have emerged as a "trust problem" due to the interaction between objective risk and subjective risk perception. Distrust in food safety is linked not only to trust in companies, but also to trust in supervisory agencies and systems as a whole, shaking the structure of social trust. Therefore, it is essential not only to strengthen technical regulations but also to restore trust through transparent information disclosure and effective risk communication to ensure quality safety.

1.4.2 Import-Oriented/Brand-Oriented Consumer Behavior

Concerns regarding the safety of food products are also manifesting as changes in consumer behavior. A prime example of this is the growing preference for imported foods and well-known brands. Cumulative experiences with domestic food safety incidents have led to a heightened perception of risk not only toward individual companies but also toward the "domestic" category as a whole.

Since the 2008 Sanlu infant formula scandal, demand for foreign brands has surged in the infant formula market, with many consumers turning to direct purchases in Hong Kong and overseas, as well as cross-border e-commerce. This shift serves as concrete evidence that distrust in food safety has translated into international purchasing behavior. Related studies have also pointed out that this incident led to a decline in consumer trust and behavioral changes in the dairy market (Pei et al., 2011; Qiao et al., 2012).

Brand preference can be understood in a similar context. As mentioned earlier, food is a typical "trust good," and its quality cannot be fully assessed before consumption. Consequently, consumers tend to use "signals", such as brand, price, and country-of-origin labels, to mitigate risk. As Akerlof's (1970) "lemons market" theory demonstrates, in markets with significant information asymmetry, trusted brands function as a substitute for quality guarantees. In the Chinese market as well, major domestic and foreign brands have built trust by emphasizing their quality control systems and the acquisition of international certifications.

Empirical studies have also confirmed that concerns about food safety tend to increase consumers' preference for imported foods and higher-priced brands. Chinese consumers consider safety, certification, brand, and country-of-origin labeling to be key factors in their food choices, and a willingness to pay for food safety attributes has also been observed (Liu et al., 2013; Ortega et al., 2011).

However, a growing preference for imported goods may also deal a blow to domestic industries and lead to market polarization. While high-income groups can mitigate risks by purchasing high-priced imported foods, low-income groups have no choice but to rely on low-priced domestic products, resulting in an imbalance in risk distribution. Thus, the issue of trust is not merely a matter of consumer behavior but is also linked to the structure of social stratification.

Overall, the growing preference for imported and branded products can be interpreted as an attempt to use market mechanisms to compensate for a lack of institutional trust in food safety. This is both a rational adaptation adopted by consumers when faced with uncertainty and an indicator of the delay in restoring institutional trust.

1.4.3 Cross-Border E-Commerce and Reliance on Digital Metrics (Including Takeout)

Trust issues related to food safety have also come to the fore in the form of the expansion of cross-border e-commerce and reliance on online reviews. In a situation where institutional trust has not been sufficiently established, consumers are increasingly turning to alternative trust mechanisms within the market to mitigate risks.

First, there is the expansion of cross-border e-commerce. Since the 2008 Sanlu infant formula scandal, demand for overseas products, particularly dairy products and health foods, has increased. This demand has been driven by Tmall Global (a cross-border e-commerce platform operated by the Alibaba Group that enables overseas brands and businesses to sell products directly to Chinese consumers without establishing physical stores in China) and JD Global (a cross-border e-commerce platform operated by the JD Group and one of the key channels for overseas brands to enter the Chinese market). Cross-border e-commerce serves as an alternative that allows consumers to bypass distrust in domestic regulatory systems and rely instead on foreign quality standards and brand credibility.

Second is the growing reliance on digital ratings. As mentioned earlier, food is a classic example of a "credence good," meaning that consumers find it difficult to verify its safety and quality directly, not only before consumption but also after. For this reason, consumers rely on online reviews, ratings, and word-of-mouth as key indicators for assessing quality.

In China, platforms such as Dianping and Meituan play a key role in visualizing evaluation information regarding restaurants and food retailers. Dianping is a leading online review site, while Meituan is a lifestyle service platform centered on food delivery. On these platforms, star ratings, the number of reviews, and user comments strongly influence consumer choice, making online reviews a critical source of information for decision-making.

Third, the rapid expansion of the food delivery market is also linked to changes in trust structures. Centered around Meituan and Ele.me, online ordering and instant delivery have become an integral part of urban life. According to the China Internet Network Information Center, the number of online food delivery users has reached hundreds of millions (China Internet Network Information Center, 2023). In the food delivery sector, because consumers cannot directly verify the hygiene conditions of physical stores, they rely heavily on platform ratings, photos, delivery speed, and certification badges. This situation reinforces a structure in which digital evaluations serve as the primary foundation for trust formation. Details will be discussed further in Chap. 7.

However, online reviews are also subject to manipulation and fake reviews, and in some cases, the very credibility of the platform itself is called into question. Therefore, while cross-border e-commerce and reliance on online reviews represent a trend toward supplementing institutional distrust with market-based and technological mechanisms, they also harbor new risks to trust.

Overall, the expansion of cross-border e-commerce and reliance on online reviews represent a phenomenon in which trust issues related to food safety have materialized at the level of consumer behavior. While this constitutes a form of "trust formation through choice" that compensates for a lack of institutional trust, improvements in institutional transparency and oversight capabilities are essential for achieving long-term stability.

1.4.4 The Differentiation Between "Institutional Trust" and "Organizational Trust"

Issues of trust regarding food safety do not manifest as a single form of "distrust," but rather as a differentiation between "institutional trust" and "organizational trust." This differentiation is crucial for understanding the structural characteristics of food safety governance in China.

First, **institutional trust** refers to trust in the government's supervisory capacity, the effectiveness of the legal system, and the transparency of information disclosure. Since the 2008 Sanlu infant formula scandal, food safety issues have come to be perceived as linked not only to corporate misconduct but also to deficiencies in the regulatory system and delays in information disclosure. Theoretical research on institutional trust identifies government fairness, competence, and transparency as core elements in the formation of trust. In China as well, the revision of the Food Safety Law and the establishment of the State Administration for Market Regulation (SAMR) are viewed as measures aimed at restoring institutional trust.

On the other hand, **organizational trust** is based on evaluations of individual companies' or brands' quality control capabilities, ethical standards, and accountability. With the development of the market economy, major domestic and foreign companies have emphasized quality assurance and traceability systems, building trust through their corporate brands. Organizational trust can serve as an alternative source of trust when institutional trust is shaken. For example, the phenomenon of consumers preferring specific well-known brands or imported goods indicates a tendency to judge trust at the corporate level rather than at the level of the institutional system as a whole (Liu et al., 2013; Ortega et al., 2011).

It is interesting to note that in China, institutional trust does not necessarily align with organizational trust. It is necessary to distinguish between general trust in the government and trust in specific policy areas, such as food safety. Research on food safety governance has also shown that while the government continues to develop institutional frameworks, consumers tend to rely on market signals, such as corporate brands, certifications, origin, and price, when making specific food choices (Liu et al., 2013, 2019).

This differentiation also affects how responsibility is attributed when food safety issues arise. In some cases, only the company is criticized when an incident occurs, while in others, institutional flaws or the supervisory responsibilities of local governments are called into question. The interaction between trust in the system and trust in companies has a complex impact on consumer behavior, brand reputation, and policy support.

In other words, the differentiation between "institutional trust" and "organizational trust" indicates that China's food safety issues are not merely a market failure, but lie at the intersection of governance structures and market structures. Ensuring food safety requires not only voluntary quality control by companies but also the strengthening of trust in the system through transparent and consistent institutional operations.

1.5 Research Objectives, Methods, and Structure of This Book (Revised Edition)

1.5.1 Research Objectives

This book aims to provide a comprehensive analysis of food safety issues in China from the perspective of trust. Food safety is not merely a matter of hygiene management or risk control; it is a complex issue closely linked to consumer risk perception, institutional governance, and social structures, including the digital economy.

In particular, this book focuses on the processes of trust formation, erosion, and reconstruction surrounding food safety issues, with the primary aim of clarifying how these processes interact with individual behavior, institutional design, and changes in social structure. To this end, it presents a three-layered trust model comprising "psychological, institutional, and social structure" as an analytical framework. The theoretical details and structure of this model are discussed systematically in Chap. 2.

1.5.2 Research Methods and Approach

This book employs a multidisciplinary research methodology that combines theoretical analysis, empirical analysis, and comparative institutional analysis. First, based on risk perception theory and trust theory, we theoretically organize the relationship between food safety and consumer trust. Through this, we present a framework that reconceptualizes food safety issues as an interrelated "risk-trust-institution" nexus. Second, using structural equation modeling (SEM) based on survey data, we empirically examine the mechanisms underlying consumer risk perception and trust formation. This allows to quantitatively elucidate the process of psychological trust formation. Third, through institutional analysis, including comparisons with Japan, we examine the relationship between the design of legal systems, administrative oversight, and risk communication, and institutional trust. This will clarify the characteristics and challenges of food safety governance in China. Furthermore, using the food delivery market and digital platforms as case studies, we will conduct a policy science analysis of the impact of sociocultural factors, such as corporate behavior, information transparency, and the evaluation economy, on trust formation.

1.5.3 The Structure of This Book and Its Correspondence with the Three-Layer Model

The book is organized into three parts corresponding to the analytical levels of the three-layered trust model (see Table 1.4).

Part I focuses on psychological trust (at the individual level) and analyzes consumers' risk perception and behavioral change, as well as their relationship with brands and the review economy. Here, we use structural equation modeling (SEM) to empirically elucidate the psychological mechanisms underlying trust formation.

Part II examines institutional trust (at the governance level) and examines the institutional design of legal systems, administrative oversight, and risk communication related to food safety. In particular, we introduce a comparative perspective between Japan and China to clarify the conditions for the formation of institutional trust and the challenges involved.

Table 1.4 The three-part structure of this book and three-layer model

Part of the book	Three-layer model	Level of analysis	Core themes	**Methodology**
Part I: Structural Analysis of Trust	Psychological trust	Individual level	Consumer risk perception, behavioral change, and their relationship with branding and the reputation economy	Structural Equation Modeling (SEM)
Part II: Institutionalization of Governance and Trust	Institutional trust	Governance level	Institutional design of legal systems, administrative regulation, and risk communication	Comparative Institutional Analysis (China–Japan Comparison)
Part III: Digital Economy, Sustainability, and Reconstruction of Trust	Social-structural trust	Socio-economic structural level	Food delivery markets, digital platforms, corporate responsibility, and sustainability	Policy Science and Social Structural Analysis

Part III focuses on socio-structural trust (at the socio-economic structural level), analyzing factors such as the takeout market, digital platforms, corporate responsibility, and sustainability. Here, we examine how the advancement of the digital economy is reshaping the relationship between food safety and trust.

1.5.4 The Positioning and Significance of This Book

As described above, by aligning the three-layered model of trust with its three-part structure, this book organically integrates theory, methodology, and empirical evidence to reframe food safety issues as a "structure of trust." Whereas previous research has been fragmented across distinct fields, this book bridges three analytical levels, psychological, institutional, and social structures, to present a comprehensive framework that views trust as a dynamic and multilayered process.

1. The international context and sustainability

Food safety is a cross-border issue, and as global supply chains expand, shortcomings in a single country's safety management can have international repercussions. Food safety is also closely linked to the United Nations Sustainable Development Goals (SDGs), specifically Goal 2 (Zero Hunger), Goal 3 (Good Health and Well-being), Goal 12 (Responsible Consumption and Production), and Goal 16 (Peace, Justice and Strong Institutions), and is positioned as the foundation of a sustainable society.

Consequently, food safety in China is not merely a domestic hygiene issue but a structural challenge linked to environmental conservation, corporate responsibility, and the reliability of international trade. Factors such as pesticide residues, heavy metal contamination, microbial risks, hygiene management during processing and distribution, and the effectiveness of oversight systems are interrelated, forming a multi-layered risk structure.

This objective risk structure influences consumers' risk perception and sense of trust, while simultaneously necessitating institutional and socio-structural trust. Focusing on this point, this book reinterprets food safety as a "structure of trust."

2. Academic significance and originality

The academic contributions of this book are organized in the following points.

The first is the theoretical integration of food safety research. Conventional research has been divided into economic analysis based on information asymmetry, psychological and sociological research focusing on risk perception, and policy research dealing with institutional design. This book presents a cross-disciplinary analysis framework by integrating these and reconstructing them as a link between "risk, trust, and system".

The second is the presentation of the "three-tier structure of trust (psychological, institutional, and social structure)". It is unique in that it takes an integrated view of psychological trust based on individual risk perception, institutional trust based on legal systems and governance, and social structural trust related to the digital economy and sustainability, and theoretically clarifies their interactions.

Third, it is a theoretical contribution to Chinese studies. By placing China's food safety issues within a generalizable theoretical framework, we theoretically explain its institutional characteristics and governance structure and present an analysis that incorporates modern factors such as digitalization and platform economy.

1.6 Policy Implications: Restructuring Food Safety Governance Through the Integration of Regulation and Trust

The analysis of this chapter reveals that the food safety problem in China is not just a accumulation of individual accidents and technical risks, but a structural issue in which "food security" and "food safety" intersect, and the latter in particular eventually emerges as a problem of "trust". In other words, food safety should be understood as an interaction of trust between institutions, companies, and consumers, and its restructuring will be the core of policy issues.

In addition, this chapter shows that while the importance of the traditional regulation-centric approach is not sufficient, it alone is not sufficient and that the development of integrated governance that incorporates trust building is essential. In other words, "regulation" and "trust" are not substitute relationships, but complementary relationships.

Specifically, first, it is necessary to standardize the supervisory system, thoroughly disclose information, and correct the incentive structure of local governments as a way to strengthen institutional trust. Second, companies need to design systems that promote the formation of autonomous trust through advanced traceability, information disclosure, and third-party certification. Third, on the consumer side, it is essential to strengthen education and risk communication to optimize risk awareness. Fourth, it is important to promote the visibility of supply chains and clarify the responsibilities of platform companies through the use of digital technology. Fifthly, it is necessary to build a safe and sustainable supply system through the redesign of food and food policies that integrate "quantity" and "quality".

Based on the above, China's food safety policy is in the process of developing into a multi-layered governance that is based on regulations and incorporates trust-building. The three-tier trust model (psychological, institutional, and social structure) presented in this chapter is an analytical framework that theoretically supports this transition and provides an effective perspective for future policy design and empirical research. Trust is not only a psychological phenomenon, but also a product of a multi-layered institutional structure.

References

Akerlof, G. A. (1970). The market for "lemons": Quality uncertainty and the market mechanism. Quarterly Journal of Economics, 84(3), 488–500. https://doi.org/10.2307/1879431

Brown, L. R. (1995). *Who will feed China? Wake-up call for a small planet*. W. W. Norton.

China Internet Network Information Center. (2023). *The 52nd Statistical Report on China's Internet Development*. China Internet Network Information Center.

Darby, M. R., & Karni, E. (1973). Free competition and the optimal amount of fraud. *The Journal of Law and Economics, 16*(1), 67–88. https://doi.org/10.1086/466756

EC-FAO Food Security Programme. (2008). *An introduction to the basic concepts of food security*. Food and Agriculture Organization of the United Nations.

Food and Agriculture Organization of the United Nations. (2022). *FAOSTAT statistical database*. Food and Agriculture Organization of the United Nations.

Food and Agriculture Organization of the United Nations, & World Health Organization. (2003). *Assuring food safety and quality: Guidelines for strengthening national food control systems*. Food and Agriculture Organization of the United Nations.

Gale, F., & Hu, D. (2012). Food safety pressures push integration in China's agricultural sector. *American Journal of Agricultural Economics, 94*(2), 483–488. https://doi.org/10.1093/ajae/aar069

Kasperson, R. E., Renn, O., Slovic, P., Brown, H. S., Emel, J., Goble, R., Kasperson, J. X., & Ratick, S. (1988). The social amplification of risk: A conceptual framework. *Risk Analysis, 8*(2), 177–187. https://doi.org/10.1111/j.1539-6924.1988.tb01168.x

Klein, J. A. (2013). Everyday approaches to food safety in Kunming. *The China Quarterly, 214*, 376–393. https://doi.org/10.1017/s0305741013000325

Kharas, H. (2017). The unprecedented expansion of the global middle class: An update.

Lam, H. M., Remais, J., Fung, M. C., Xu, L., & Sun, S. S. (2013). Food supply and food safety issues in China. *The Lancet, 381*(9882), 2044–2053. https://doi.org/10.1016/S0140-6736(13)60776-X

Liu, P., Ma, L., & Bai, Y. (2019). Food safety governance in China: From supervision to coregulation. *Food Science & Nutrition*, 7(12), 4127–4139. https://doi.org/10.1002/fsn3.1281

Liu, R., Pieniak, Z., & Verbeke, W. (2013). Consumers' attitudes and behaviour towards safe food in China: A review. *Food Control, 33*(1), 93–104. https://doi.org/10.1016/j.foodcont.2013.01.051

Ministry of Environmental Protection of the People's Republic of China, & Ministry of Land and Resources of the People's Republic of China. (2014). *Report on the national general survey of soil contamination*.

National Bureau of Statistics of China. (2023). *China statistical yearbook 2023*. China Statistics Press.

National People's Congress of the People's Republic of China. (2015). *Food Safety Law of the People's Republic of China*.

OECD, & Food and Agriculture Organization of the United Nations. (2023). *OECD-FAO agricultural outlook 2023–2032*. OECD Publishing. https://doi.org/10.1787/08801ab7-en

Ortega, D. L., Wang, H. H., Wu, L., & Olynk, N. J. (2011). Modeling heterogeneity in consumer preferences for select food safety attributes in China. *Food Policy, 36*(2), 318–324. https://doi.org/10.1016/j.foodpol.2010.11.030

Pei, X., Tandon, A., Alldrick, A., Giorgi, L., Huang, W., & Yang, R. (2011). The China melamine milk scandal and its implications for food safety regulation. *Food Policy, 36*(3), 412–420. https://doi.org/10.1016/j.foodpol.2011.03.008

Qiao, G., Guo, T., & Klein, K. K. (2012). Melamine and other food safety and health scares in China: Comparing households with and without young children. *Food Control, 26*(2), 378–386. https://doi.org/10.1016/j.foodcont.2012.01.045

State Council of the People's Republic of China. (2016). *Healthy China 2030 Planning Outline*. State Council of the People's Republic of China.

State Council of the People's Republic of China. (2019). *Food safety in China*. The State Council Information Office of the People's Republic of China.

The Paper. (2022, November 30). Zui shou guanzhu de shi da jiaodian wenti: yiqing zai deng bangshou, jingji xiaxing qiandong minxin [Top ten issues of greatest public concern: COVID-19 prevention and control ranks first again while economic slowdown concerns the public].

World Bank. (2021). *China—Food Safety Improvement Project*. World Bank.

World Bank. (2022). *The greening of China's agriculture: A policy brief*. World Bank.

World Health Organization. (2009). *Toxicological and health aspects of melamine and cyanuric acid: Report of a WHO expert meeting in collaboration with FAO, supported by Health Canada*. World Health Organization.

World Health Organization. (2015). *WHO estimates of the global burden of foodborne diseases: Foodborne disease burden epidemiology reference group 2007–2015*. World Health Organization.

World Health Organization. (2022). *Noncommunicable diseases country profiles 2022*. World Health Organization.

Xiaokang Magazine. (2005, November 24). *Yingxiang xiaokang jincheng de shida jiaodian wenti: Fubai wenti shengzhi disan* [Top ten focal issues affecting the process of building a well-off society: Corruption rises to third place]. Sina News.

Xiaokang Magazine. (2007, December 3). *Quanmian xiaokang zuishou guanzhu de shidajiaodian* [Top ten focal issues most concerned about in building a well-off society]. Sina News.

Chapter 2
Theoretical Foundations of Food Safety: Risk Perception and the Construction of a Three-Layer Trust Model

Weisheng Zhou, Fengyang Wang, and Xuepeng Qian

2.1 Food-Related Risks and Risk Perception Theory

Food safety refers to ensuring, in scientific and objective terms, that food does not cause adverse effects on human health. This is achieved through processes of risk assessment and risk management. In contrast, "assurance" goes beyond the mere fact of scientific safety and refers to a psychological state in which consumers accept and feel confident about the safety of food. In other words, assurance emerges only when both objective safety and trust in that safety are established. In this regard, research on food risk has emphasized the importance of the interaction between scientific risk assessment and consumer trust (Slovic, 1987; Frewer et al., 1995). Since trust is formed through transparent information provision and the evaluation of safety measures implemented by governments, firms, and regulatory institutions, both the assurance of food safety and the construction of consumer trust are essential for achieving consumer assurance. Thus, while scientific management of food safety provides the foundation for assurance, trust in institutions and relevant actors is equally indispensable for its formation (de Jonge et al., 2004).

In analyzing how assurance can be achieved, it is necessary to first examine the causes of consumer anxiety, given that assurance is essentially the inverse of anxiety. Existing studies suggest that consumer anxiety regarding food safety is primarily determined by risk perception and trust in relevant actors (Slovic, 1987; Lobb et al.,

W. Zhou (✉)
College of Policy Science, Ritsumeikan University, Kyoto, Japan
e-mail: zhou@sps.ritsumei.ac.jp

F. Wang
Xi'an Peihua University, Xi'an, China

X. Qian
Graduate School of Global Environmental Studies, Sophia University, Tokyo, Japan
e-mail: qianxp@sophia.ac.jp

W. Zhou (ed.), *Food Safety in China*, https://doi.org/10.1007/978-981-92-3741-8_2

2007). In other words, the questions of why consumers fail to feel assured (i.e., why anxiety arises) and how assurance can be achieved (i.e., how anxiety can be reduced) are closely related to theories of risk perception and trust.

As discussed in Chap. 1, food safety issues in China should not be understood merely as problems of hygiene management or risk control, but rather as a "structure of trust" that is closely linked to consumers' risk perception, institutional governance, and broader social structures, including the digital economy. It was also pointed out that existing studies remain fragmented across disciplines such as economics, sociology, and policy studies, and that there is a need for an integrated theoretical framework.

Based on these considerations, this chapter systematically examines the theoretical foundations of food safety, assurance, and trust, and develops the three-layered trust model that constitutes the core of this book. This section reviews food-related risks and risk perception theory, clarifying the relationship between objective risk and subjective perception, as well as the specific characteristics of the risk perception gap in the context of food. Section 2.2 then examines the definition and formation process of trust, and analyzes the relationships among food safety, assurance, and trust in a functionally differentiated society. Furthermore, Sect. 2.3 identifies the limitations of existing theories of risk, trust, and institutions, highlighting the theoretical gaps arising from their insufficient integration. Building on this analysis, Sect. 2.4 presents the central theoretical contribution of this book: the three-layered trust model. In this model, three dimensions are defined: psychological trust (individual level), institutional trust (governance level), and structural trust (socio-economic level), along with their interdependent relationships. Finally, Sect. 2.5 clarifies the correspondence between the three-layered model and the overall analytical design of the book and situates the empirical and institutional analyses developed in Parts I through III.

2.1.1 Clarifying the Definition of Risk Cognition: Risk Perception and Risk Cognition

In research on the psychological evaluation of risk, the concept of *risk cognition* is frequently employed, while related terms such as *risk perception* and *cognitive structure* are also used. Conceptually, it is necessary to distinguish and organize these terms. *Risk cognition* can be understood as the cognitive process through which individuals estimate and evaluate the magnitude and characteristics of risk—that is, a form of psychological evaluation involving structured information processing (Tversky & Kahneman, 1974). This individual decision-making process is not static; rather, it integrates evaluations of various social factors and is continuously updated depending on the context, thereby constituting a complex process of information processing.

By contrast, *risk perception* refers to the intuitive and subjective evaluation that arises when individuals encounter risk-related information. It is conceptualized as a psychological process strongly influenced by heuristics and cognitive biases (Slovic, 1987; Tversky & Kahneman, 1974). In this sense, risk perception is relatively immediate and intuitive, whereas risk cognition encompasses a broader and more integrative cognitive process.

The focus of this book is on a comprehensive psychological evaluation that includes both intuitive judgments and deliberative reasoning. Specifically, it adopts an integrated perspective in which risk cognition—understood as a relatively stable, cognitively processed evaluation—serves as the foundation, while also incorporating elements of risk perception. For analytical convenience, therefore, this study uses the term *risk perception* to denote the overall psychological evaluation of food-related risks by consumers.

2.1.2 Risk and Risk Cognition

Risk has been studied across various academic disciplines, and its definition differs depending on the field; nevertheless, efforts have been made to establish general definitions. In economics and decision theory, risk is understood as uncertainty associated with gains or losses, that is, a probabilistically measurable phenomenon (Knight, 1921). More generally, risk is often defined as the combination of the probability of an undesirable event and the severity of its consequences (Kaplan & Garrick, 1981). This understanding forms the basis of the concept of risk in many contemporary fields.

In the domain of food safety, the concept of risk follows this general framework. Internationally, the Codex Alimentarius Commission (CAC) defines risk as "a function of the probability of an adverse health effect and the severity of that effect, consequential to a hazard(s) in food" (Codex Alimentarius Commission, 2006). According to this definition, food risk is a statistical concept determined by factors such as (1) the probability of the presence of hazards, (2) the level of food consumption (exposure), and (3) the susceptibility of individuals.

Furthermore, in the field of food safety, a clear distinction is made between *risk* and *hazard*. A hazard refers to a biological, chemical, or physical agent in food with the potential to cause adverse health effects, whereas risk refers to the likelihood and severity of such effects actually occurring (Codex Alimentarius Commission, 2006). Since all foods inherently contain some level of hazard, it is impossible to reduce risk to zero. Accordingly, food safety is understood as the management of risk to a socially acceptable level.

In addition, the framework of *risk analysis* is widely adopted in modern food safety management. This framework consists of three components: (1) risk assessment as a scientific evaluation, (2) risk management as a policy decision-making process, and (3) risk communication as the exchange of information among stakeholders (FAO & WHO, 1995). This framework provides an essential basis for understanding food safety not merely as a technical issue, but as a process of social decision-making.

Figure 2.1 illustrates the relationship between the intake (exposure level) of hazards in food and their effects on the human body, commonly referred to as the dose–response relationship. The horizontal axis represents the level of intake, while the vertical axis represents the degree of health impact. At low levels of intake, no adverse health effects are observed, remaining within the range of the no-effect level or the maximum no-effect level. However, once a certain threshold is exceeded, biological responses begin to occur, and with further increases in intake, clear health effects such as toxicity may emerge. Ultimately, excessive intake may lead to severe outcomes, including death. Thus, this figure demonstrates that risk is not determined solely by the intrinsic harmfulness of a hazard, but increases progressively depending on the level of exposure (intake).

The concept of risk cognition was first introduced into the field of marketing by Bauer (1960) and has since developed into an important analytical framework in consumer behavior research. In particular, risk cognition has been conceptualized as a two-dimensional construct consisting of uncertainty and the severity of consequences (Cunningham, 1967). In other words, risk cognition can be understood as the integration of a subjective probability that an undesirable event will occur and an evaluation of the seriousness of its consequences.

However, human risk cognition is not necessarily based on rational probability calculations; rather, it is strongly influenced by intuitive judgments relying on heuristics. For example, heuristics such as availability and representativeness systematically distort the perceived magnitude of risk (Tversky & Kahneman, 1974). Moreover, risk cognition has been shown to be structured along psychological dimensions

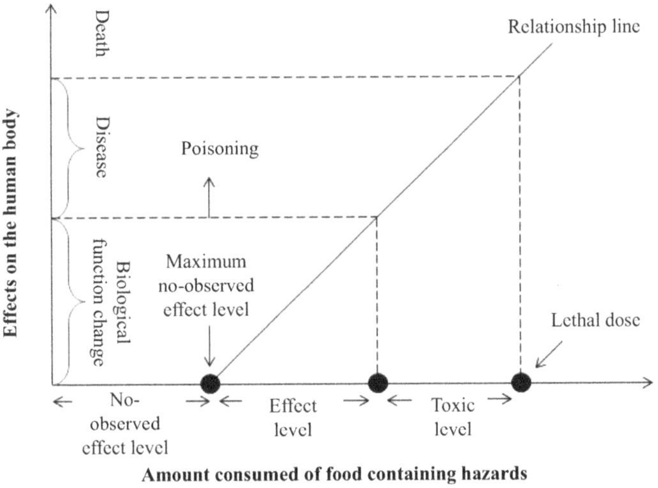

Fig. 2.1 Relationship between intake (exposure) to foodborne hazards and health effects. *Note* The maximum no-effect level does not imply a complete absence of biological impact, but rather the highest level at which no observable adverse biological effects are detected. *Source* Prepared by the author based on various sources

such as *dread* and *unknown*, which are considered fundamental components of how risks are perceived (Slovic, 1987). These cognitive characteristics vary depending on individuals' knowledge and experience and directly influence consumer attitudes and decision-making (Mitchell, 1999).

In the context of food, risk cognition exhibits additional complexity due to the intrinsic characteristics of food itself. On the one hand, food is indispensable for human survival; on the other hand, risk can never be completely eliminated, and safety must therefore be understood as a relative concept. Furthermore, food-related risks vary depending on the level of intake (exposure) and individual differences in susceptibility, implying that a zero-risk condition does not exist.

In addition, food reaches consumers through multiple stages—production, processing, and distribution—many of which are not directly observable to consumers. This lack of transparency creates information asymmetry, which in turn affects risk cognition. Moreover, differences in dietary habits and food culture mean that the acceptance of risk varies across countries and regions, even for the same food product. Thus, food risk is not only a matter of scientific evaluation but is also closely linked to social and cultural factors.

These characteristics are systematically reflected in international food safety frameworks. Food risk is defined as "a function of the probability of an adverse health effect and the severity of that effect," and is addressed within the framework of risk analysis, which consists of risk assessment, risk management, and risk communication (Codex Alimentarius Commission, 2006). Furthermore, the importance of understanding risk, taking into account the diversity and uncertainty of foods as well as the limitations of scientific knowledge, is also emphasized in international risk analysis frameworks (Codex Alimentarius Commission, 2006; FAO & WHO, 1995).

In summary, while food-related risk cognition is grounded in general theories of risk perception, it exhibits a more complex, multi-layered structure that is highly dependent on the specific characteristics of food and its broader social context.

2.1.3 The Gap in Food Risk Perception

The process by which individuals perceive and evaluate risk is referred to as risk cognition. While expert judgments are typically based on scientific and objective indicators, laypeople tend to rely more on intuitive and subjective evaluations (Fischhoff et al., 1978). As a result, objective risk and subjectively perceived risk do not necessarily coincide, and this divergence is commonly described as the *risk perception gap*.

Studies on subjective risk perception have shown that individuals systematically misjudge the probability of various events associated with mortality risk. Specifically, low-probability events tend to be overestimated, whereas high-probability events are underestimated (Lichtenstein et al., 1978). For example, risks such as nuclear accidents or natural disasters—characterized by sudden occurrence and high salience—are often overestimated, whereas more familiar risks, such as traffic accidents, are underestimated.

Furthermore, risk perception is strongly influenced by psychological dimensions such as *dread* and *unknown*, which are key factors underlying differences between expert and public risk evaluations (Slovic, 1987). In other words, while experts assess risks based on statistical probabilities, consumers tend to rely on emotions and intuitive judgments, leading to systematic discrepancies between the two.

A similar gap exists in the domain of food safety. In particular, risks associated with artificial factors, such as food additives and pesticides, tend to be overestimated, whereas risks associated with natural substances are often underestimated (Slovic, 1987). This gap in risk perception constitutes a major challenge for food safety policy and risk communication.

Table 2.1 presents mortality risks in everyday life. As shown, there is a systematic divergence between objective risk levels and subjective risk perception. In particular, voluntary risks tend to be underestimated, while involuntary risks are often overestimated (Slovic, 1987). Such asymmetry in risk perception is not merely an individual cognitive bias but plays a critical role in shaping public anxiety and distrust regarding food safety.

In the context of food, risks are often perceived as involuntary, since consumers typically have limited control over production and distribution processes. This perception contributes to the overestimation of risk and the amplification of anxiety. As a consequence, even when objectively assessed risks are low, subjective risk perceptions may remain elevated, thereby undermining trust in food safety.

A divergence between objective risk and perceived risk is also well recognized in the field of food safety. In the context of food, the process of estimating the magnitude of risk—both qualitatively and quantitatively (i.e., risk assessment)—is

Table 2.1 Annual mortality risks in everyday life

Category	Activity/Event	Annual mortality risk
Voluntary	Automobile driving	~1×10^{-4}
Voluntary	Alcohol consumption (long-term effects)	~1×10^{-5}
Voluntary	Motorcycle riding	~1×10^{-3}
Involuntary (Natural hazards)	Flood	~1×10^{-6}
Involuntary (Natural hazards)	Lightning	~1×10^{-7}
Involuntary (Natural hazards)	Tornado	~1×10^{-7}
Involuntary (Environmental exposure)		

Note The values represent approximate annual mortality risks per individual. Prepared by the author based on various statistical sources. The comparison between voluntary actions and involuntary risks illustrates gaps in risk perception conducted objectively by governments, scientists, and specialized institutions based on historical data and scientific evidence (FAO & WHO, 1995). The results derived from such processes are regarded as *objective risk*, representing the actual level of risk.

However, the risks that individuals perceive with regard to food—namely, *subjective risk*—are strongly influenced by psychological and social factors and therefore often do not correspond to scientifically assessed objective risk. In the process of forming risk perceptions related to food safety, individual knowledge, experience, values, and broader social contexts all shape how risk is evaluated, leading to systematic tendencies toward overestimation or underestimation (Slovic, 1987).

Figure 2.2 presents a conceptual framework illustrating how risk perceptions related to food are formed and why they lead to consumer anxiety and distrust. The figure conceptualizes food risk from two dimensions—objective risk and subjective risk—and explains the mechanism through which their interaction generates a perception gap.

First, *objective risk* is derived from scientific risk assessment based on empirical data and expert knowledge, and is composed of elements such as the probability of occurrence and the severity of potential impacts (FAO & WHO, 1995). Second, *subjective risk* refers to the psychological risk perceived by consumers, which is strongly influenced by factors such as uncertainty, controllability, and the comprehensibility of information. In particular, risks that are perceived as unknown, difficult to understand, or beyond individual control tend to be overestimated, whereas familiar risks or those associated with clear benefits are more likely to be underestimated (Slovic, 1987).

The discrepancy between objective and subjective risk manifests as a cognitive misalignment, commonly referred to as the *risk perception gap*. This gap represents the divergence between "actual risk," as scientifically assessed, and "perceived risk,"

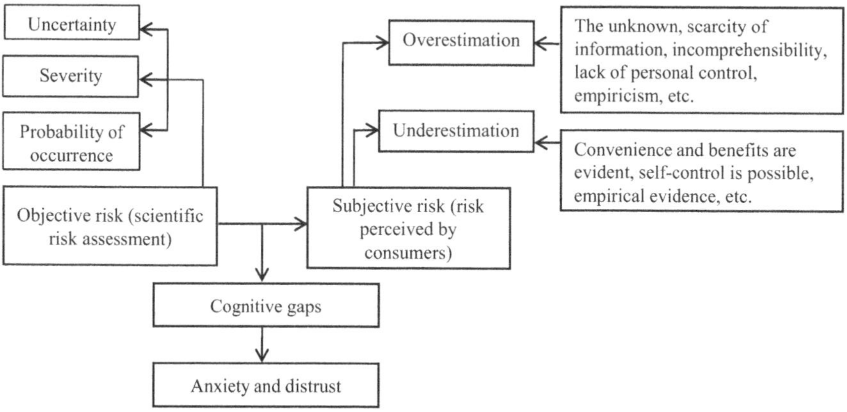

Fig. 2.2 The risk perception gap in the structure of food risk cognition (conceptual diagram). *Source* Developed by the author

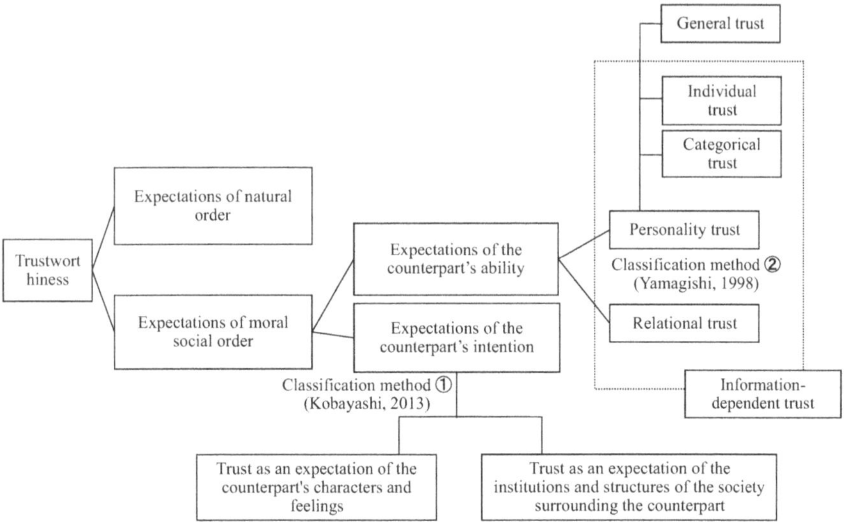

Fig. 2.3 Theoretical classification of trust. *Note* Compiled by author from Barber (1983), Luhmann (1979), and Yamagishi (1998)

as experienced by consumers, and constitutes a key factor underlying anxiety and distrust toward food safety (Frewer et al., 1995). Moreover, as this perception gap widens, consumers are more likely to develop heightened anxiety and distrust toward food and food safety systems. In particular, insufficient information provision and inadequate risk communication tend to amplify subjective risk, leading to increased public concern. As a consequence, excessive safety-related costs and broader social losses may arise.

Therefore, in the field of food safety, it is essential for governments, industry actors, and experts to collaborate in narrowing the gap between objective and subjective risk through effective risk communication (FAO & WHO, 1995).

2.2 Concept and Structure of Trust

2.2.1 Definition and Characteristics of Trust in This Study

Trust is an essential element for the effective functioning of organizations and societies, and is widely recognized as a fundamental basis of social relationships, including political and economic interactions. Research on trust has been developed across multiple disciplines, resulting in a concept that is inherently multidimensional and polysemous. In a broad sense, trust can be understood as expectations toward others or society. In this regard, Barber (1983) defines trust as "expectations of the persistence of natural and moral social order." Since this study focuses on trust in the context of human interaction, it emphasizes expectations toward moral social order.

Regarding social trust, Yamagishi (1998) distinguishes expectations toward others into two types: expectations of competence and expectations of intention. He particularly identifies the latter as the core of trust. In other words, trust refers to the expectation that, under conditions of uncertainty, others will not behave opportunistically or exploitatively. Furthermore, Rousseau et al. (1998) define trust as "a psychological state comprising the intention to accept vulnerability based upon positive expectations of the intentions or behavior of another." This definition clearly indicates that trust is inherently grounded in uncertainty and risk. In addition, trust can be categorized into cognitive trust and affective trust. Cognitive trust is a rational evaluation based on knowledge of another party's competence and reliability, whereas affective trust is formed through emotional bonds and relational attachment (McAllister, 1995). These two types of trust complement each other and jointly function within social relationships.

Based on these discussions, three essential characteristics of trust can be identified. First, trust presupposes the existence of uncertainty and risk. Second, trust is a voluntary choice involving positive expectations regarding another's behavior. Third, trust is shaped by both cognitive and affective factors. Accordingly, this study defines trust as "a positive expectation regarding the competence and intentions of another party under conditions of social uncertainty, accompanied by an evaluation that the other party will fulfill those expectations."

2.2.2 Structure of Trust and the Process of Trust Formation

As discussed above, this study defines trust as an evaluation that another party will respond to one's expectations. Previous research has demonstrated that trust is not a unitary concept but rather possesses a multi-layered structure, and that its formation processes are diverse (Rousseau et al., 1998).

2.2.2.1 Structure of Trust

Trust can first be divided into expectations regarding another party's competence and expectations regarding their intentions. In particular, the latter—expectations that the other party will not behave opportunistically under conditions of uncertainty—constitutes the core of trust (Yamagishi, 1998). Furthermore, trust as an expectation of intention can be categorized according to its basis of formation. First, trust may be based on direct relationships and experiences with a specific individual, relying on interpersonal interactions and evaluations. Second, trust may be formed based on indirect information such as reputation or attributes (e.g., profession or organizational affiliation), which involve judgments based on social categories or labels. Since both types rely on specific information about the other party, they can be classified as *information-based trust*.

In contrast, even in the absence of specific information about others, individuals may generally expect others to behave cooperatively. This form of trust is referred to as *generalized trust* and functions as a foundation for social interaction (Yamagishi, 1998). Thus, trust can be distinguished between that based on specific information and that based on generalized expectations toward others.

Importantly, trust is not limited to the interpersonal level but is also formed as expectations toward institutions and social structures. In this sense, trust can be divided into expectations toward individual actors (e.g., their intentions and character) and expectations toward institutional and social systems. The latter is supported by legal systems, regulations, and social norms, which enhance the predictability of others' behavior and thereby enable trust (Luhmann, 1979).

For example, when individuals have prior relationships, trust is formed based on personal knowledge of the other party's character and past behavior. In contrast, even in transactions between strangers, trust can be established through institutional mechanisms such as contracts and sanctions, which ensure that opportunistic behavior will be penalized. Thus, trust is supported by both interpersonal relationships and institutional frameworks, forming a multi-layered structure.

As such, the structure of trust emerges from the interaction of individual cognition and experience, social information, and institutional environments, and serves as the theoretical foundation for the multi-layered trust model proposed in this book.

2.2.2.2 Trust Formation Process: The Interaction Between Affective and Cognitive Trust

Based on prior studies, the process of trust formation can be systematically described. Trust can be broadly classified into two types according to its underlying basis: *cognitive trust* and *affective trust*. Cognitive trust refers to trust formed through an evaluation, based on knowledge and experience, that another party will act as expected, particularly in terms of competence and reliability. In contrast, affective trust is formed through long-term interactions and emotional bonds, and is supported by the accumulation of empathy and relational ties (McAllister, 1995).

The formation of affective trust is closely related to processes of meaning-making and reflexive evaluation based on individual experience. Giddens (1979) argues that human actions are continuously reconstructed through reflective monitoring of past experiences. Furthermore, Giddens (1991) introduces the concept of *reflexivity*, whereby individuals make decisions by continuously referring to and reassessing their own experiences and value systems. From this perspective, individuals interpret external information in light of consistency with their past experiences and internal value frameworks. When such consistency is recognized, it generates a sense of empathy, and the accumulation of this empathy leads to the formation of affective trust.

Moreover, affective trust is not merely a transient emotional response but a relational resource that is reinforced through repeated interactions. As mutual understanding and shared values deepen, the psychological distance between actors decreases, and trust becomes more stable and resilient. This process can be observed not only in interpersonal relationships but also in relationships between consumers and firms, as well as between citizens and institutions.

In contrast, cognitive trust is formed through processes of information processing and evaluation regarding another party. Mayer et al. (1995) identify three core components of trust: *ability*, *benevolence*, and *integrity*. Based on this framework, cognitive trust can be understood as consisting of two interrelated processes: (1) a *trait inference process*, in which individuals infer the characteristics of another party (e.g., competence, goodwill, and integrity) based on past interactions and available information; and (2) an *outcome prediction process*, in which individuals use these inferred traits to predict future behavior. In this sense, cognitive trust is formed as the result of rational evaluation grounded in accumulated experience and information.

In contemporary consumer society, the processes of production and service provision have become increasingly complex, and the invisibility of supply chains, along with growing information asymmetry, has made it more difficult for consumers to rely solely on direct experience. As a result, cognitive trust must often be formed through indirect sources such as media information, expert evaluations, and third-party certifications. In this context, continuous and transparent information provision and communication by firms and public institutions play a crucial role in fostering cognitive trust.

Importantly, affective trust and cognitive trust interact in a mutually reinforcing manner. Affective trust, grounded in empathy, encourages individuals to interpret information about others in a more favorable way, thereby facilitating the formation of cognitive trust. Conversely, the accumulation of cognitive evaluations regarding another party's competence and integrity strengthens emotional bonds, reinforcing affective trust. Thus, the two forms of trust exhibit a complementary and cyclical relationship that underpins the development of sustained trust (McAllister, 1995).

As illustrated in Fig. 2.4, trust formation is a dynamic process driven by the interaction between affective and cognitive dimensions, involving the combined influence of individual experience, social information, and institutional context. This understanding of trust formation constitutes a key theoretical foundation for the multi-layered trust model proposed in this book.

2.2.3 Existing Theories of Trust in Food Safety

The increasing complexity of food production, processing, distribution, and consumption has made it more difficult for stakeholders across the entire food chain—including consumers—to obtain and explain information related to safety and reliability. In particular, the occurrence of food incidents and corporate misconduct in recent decades has heightened awareness that food risk is closely embedded in

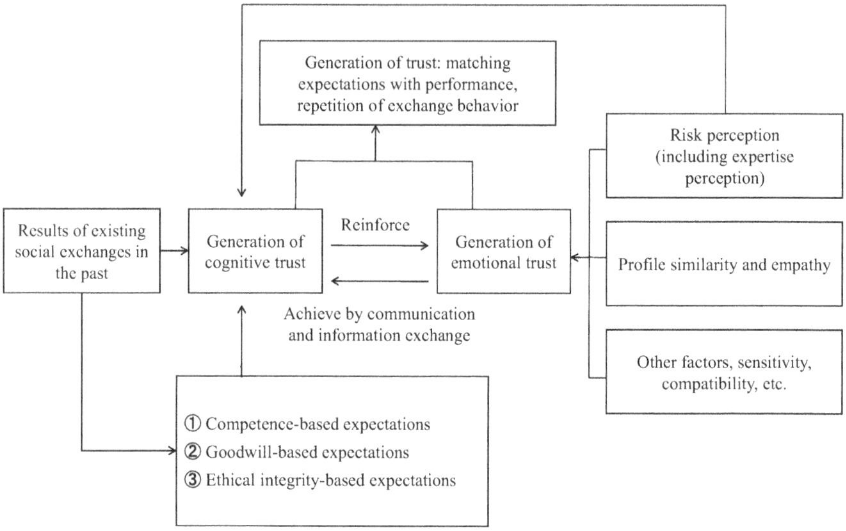

Fig. 2.4 Trust formation process: the interaction between affective and cognitive trust (conceptual diagram). *Note* Developed by the author based on Mayer et al. (1995) and McAllister (1995)

everyday life, thereby increasing consumer concerns about food safety and the food system as a whole. Against this backdrop, trust in food has emerged as a significant topic of international research (Frewer et al., 1995).

2.2.3.1 Definition and Structure of Food Safety Trust

Building on the conceptualization of trust discussed above, consumer trust in food safety can be understood as the integration of expectations regarding both the competence and the intentions of actors involved in the food system. These actors include producers, processors, distributors, retailers, as well as governmental authorities and scientific experts. In this study, such trust is defined as *food safety trust*.

The key components of food safety trust include, first, *competence* (technical capability and expertise), second, *integrity* (the absence of fraudulent behavior such as mislabeling), and third, *benevolence* (a willingness to consider consumer welfare). These elements are also identified as core dimensions of organizational trust in existing literature (Mayer et al., 1995). In the context of food safety, additional factors such as *transparency* (the openness of information disclosure) and *consistency* (the stability and reliability of behavior over time) are particularly important.

Moreover, food safety trust is not determined solely by evaluations of competence but is also strongly influenced by the alignment of values between consumers and risk management actors. When regulatory institutions adopt a consumer-oriented perspective—demonstrating what they prioritize and how they address problems—and when consumers can empathize with these approaches, trust in such institutions is enhanced. This notion of *shared values* has been identified as a critical factor in trust formation within the literature on risk communication (Frewer et al., 1995).

In addition, food safety trust is shaped by both internal and external factors. Internal factors include consumers' knowledge, experience, values, and risk perceptions, whereas external factors encompass governmental responses, corporate behavior, media coverage, technological conditions, and institutional frameworks. These factors interact with one another to influence the process of trust formation.

Furthermore, individual attributes and psychological tendencies also play an important role in shaping food safety trust. de Jonge et al. (2008) analyze consumer trust in food safety in terms of *optimism* and *pessimism*, demonstrating that these attitudes function as key mediating variables in the formation of trust. In other words, even under the same informational environment, the process of trust formation can vary significantly depending on individual psychological dispositions.

Taken together, existing studies suggest that food safety trust is a multidimensional construct shaped by the interaction of cognitive evaluations (such as competence, integrity, and value alignment), individual psychological factors, and broader social and environmental conditions. In this book, these relationships, along with their implications for consumer behavior, are systematically organized and presented in Fig. 2.5.

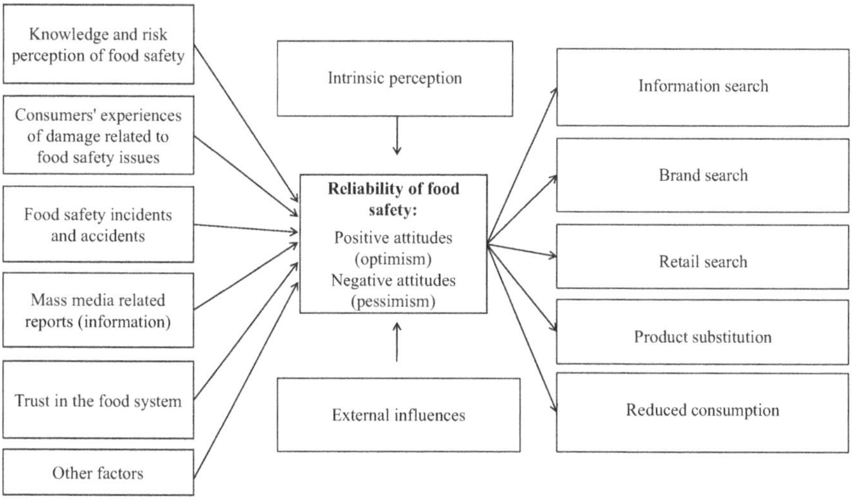

Fig. 2.5 Structural model of food safety trust (conceptual diagram). *Note* The "food system" here refers to the entire set of actors involved in food production, processing, distribution, and consumption, as well as the regulatory and administrative frameworks related to food safety, including legal systems, rules, norms, and standards. 2008 *Source* Adapted and extended by the author based on de Jonge et al. ()

2.2.3.2 Building Food Safety Trust: Trust Formation Through Risk Communication

As discussed above, from the perspective of affective and cognitive trust, the formation of food safety trust depends critically on the generation of empathy and the alignment between expectations and actual outcomes. Therefore, it is not sufficient to enhance the technical and objective safety of food alone; in order to build food safety trust, it is essential to strengthen communication, including the provision and dissemination of food-related information, as well as to promote food education that enhances scientific understanding of food safety. Through these efforts, consumers' knowledge and risk perception can be improved, thereby fostering understanding of and empathy toward the food system.

Within the framework of risk analysis, risk communication is defined as an interactive process of exchanging information and opinions among individuals, groups, and organizations concerned with risk. Its objectives include enhancing understanding of risk, narrowing the risk perception gap, and building trust among stakeholders (FAO & WHO, 1995). In other words, to build consumer trust in the food system, it is necessary for diverse actors—such as government agencies, firms, scientific experts, and the media—to collaborate in promoting appropriate and continuous risk communication.

In Western countries, since the late twentieth century, food safety crises such as bovine spongiform encephalopathy (BSE) and dioxin contamination have led to a rapid increase in public concern and a significant erosion of trust in food safety governance. In response, the importance of risk communication for restoring public trust has been widely recognized, and its approaches have been reexamined. It has been pointed out that the divergence in risk perception is often rooted in insufficient or inadequate information provision, and that effective risk communication is essential for reducing this gap and rebuilding trust (Frewer et al., 1995).

In particular, within food safety policies and measures, the communication of risk-related information to consumers is of critical importance. Transparent information provision by governments, firms, and experts directly influences consumers' risk perceptions and contributes to enhancing their understanding of and trust in food safety and related policies. Moreover, when risk information is appropriately communicated, excessive public concern can be mitigated, and more rational risk perceptions can be fostered (Slovic, 1987).

In summary, the formation of trust in food safety depends not merely on the provision of information, but on the quality and continuity of risk communication that simultaneously engages both affective and cognitive dimensions of trust.

2.2.3.3 Food Safety, Trust, and Assurance in a Functionally Differentiated Society

In modern societies characterized by advanced specialization and division of labor, the increasing complexity of the food system—from agricultural production to processing, distribution, retail, and consumption—has made it difficult for stakeholders, including consumers, to obtain and interpret safety-related information about food. This structural condition amplifies uncertainty regarding food safety and creates an environment in which consumer anxiety is easily generated. In particular, contemporary society is often described as a "risk society," in which risks are socially produced, distributed, and shared, and food safety issues represent a typical example of such dynamics (Beck, 1992).

The expansion of such anxiety extends beyond individual psychological responses and manifests as broader social losses. Through repeated exposure to food-related incidents and crises, consumers tend to learn that "trust itself entails risk." As a result, when firms pursue self-interest at the expense of integrity, the erosion of trust can trigger a cascading expansion of social damage. This problem is particularly severe because food safety incidents occurring in a specific region can have far-reaching impacts beyond their immediate locality, leading to widespread diffusion of consumer anxiety across society.

Moreover, in the process of risk communication and dissemination, risks are often amplified beyond their actual health impacts, further intensifying public concern. In other words, risk perception frequently diverges from objective risk, and is often overestimated due to social and psychological factors (Slovic, 1987). Consequently, distrust may extend not only to the firms directly responsible for incidents but also to unrelated firms and products, negatively affecting the entire market. This situation gives rise to a structural condition akin to "bad money driving out good," reflecting a systemic breakdown of trust.

Therefore, in contemporary food safety management systems, it is essential not only to enhance the objective safety of food throughout the entire chain from production to consumption, but also to ensure consumers' subjective sense of assurance. In this regard, it is necessary to integrate "objective safety" and "subjective assurance," or "technical safety" and "social assurance," within a unified analytical framework.

Furthermore, to secure consumer assurance, it is necessary to establish and enforce appropriate standards, regulations, and management systems to improve technical safety, while also clearly and transparently communicating these measures to consumers. This represents a core function of risk communication. In this process, trust serves as an indispensable mediating factor.

In other words, consumers' sense of assurance regarding food safety depends fundamentally on their trust in the food safety management system. Trust thus functions as the key mechanism linking food safety and consumer assurance (Luhmann, 1979). This relationship can be expressed as follows: *assurance is achieved through the combination of safety and trust*. Accordingly, to address the anxiety revealed by food safety issues and to ensure consumer assurance, it is essential not only to enhance technical safety but also to build trust in the food system as a whole (see Fig. 2.6).

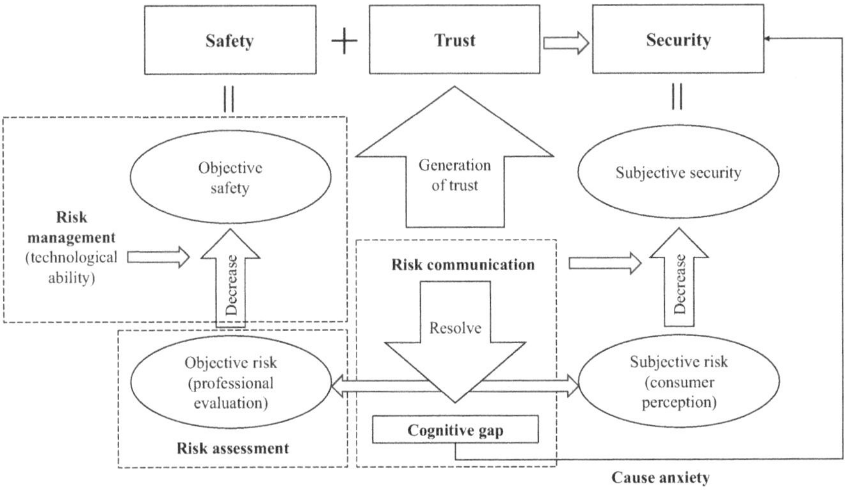

Fig. 2.6 Relationship between safety, trust, and assurance. *Note 1* "Ensuring safety + building trust → achieving assurance" indicates that ensuring safety alone is insufficient; consumer assurance is achieved only when trust is also established. *Note 2* The dashed box represents the three components of risk analysis: risk assessment, risk management, and risk communication

2.3 Review of Existing Studies on Food Safety Issues in China and Theoretical Challenges

In examining consumer anxiety regarding food safety in China, it is essential to analyze the specific characteristics of China's food safety issues and to assess the applicability of existing theoretical frameworks.

First, rapid economic growth, demographic changes, and the diversification of food consumption have led to the substantial expansion of the food industry, including food manufacturing, distribution, and the food service sector. The scale of agriculture and related industries in China has increased significantly—from 578 billion yuan in 1998 (approximately USD 70 billion at an exchange rate of 1 USD $\approx$ 8.28 yuan) to 10.89 trillion yuan in 2014 (approximately USD 1.77 trillion at an average exchange rate of 1 USD $\approx$ 6.14 yuan). According to recent statistics, the total value added of agriculture and related industries reached approximately 19.85 trillion yuan in 2023 (about USD 2.8 trillion at an average exchange rate of 1 USD $\approx$ 7.08 yuan), indicating a continued expansion trend (National Bureau of Statistics of China, 2024). As the division of labor within the food industry has deepened, consumers have become increasingly aware of and concerned with safety, assurance, and trust in the food safety system.

Second, the fundamental perspective of this study is that, although China has made serious efforts to improve food safety—enhancing regulatory standards and overall product quality—consumer anxiety has persisted. In addressing the causes of this phenomenon and exploring potential solutions, it is crucial to adopt a theoretical approach that integrates risk perception structures and trust formation mechanisms. It should be emphasized that focusing on assurance and trust does not imply neglecting the objective safety of food, nor does it suggest that all safety issues have been fully resolved. Rather, improving psychological trust among consumers is an indispensable dimension of advancing food safety in China.

Based on these considerations, existing studies on food safety issues in China can be broadly categorized into three major approaches:

1. economic studies (information asymmetry and market failure),
2. sociological studies (risk society theory and trust theory), and
3. policy studies (legal systems and administrative reforms).

The following sections review the achievements and limitations of these three research streams and clarify the theoretical challenges in the Chinese context.

2.3.1 Economic Approaches: Information Asymmetry and Market Failure

Economic studies on food safety issues in China have primarily developed within the theoretical framework of information asymmetry and market failure. A key theoretical foundation is the argument that, in markets where quality is difficult to observe ex ante, information gaps between producers and consumers can undermine transaction efficiency and reduce social welfare. Akerlof's (1970) "market for lemons" model demonstrates how quality uncertainty can lead to adverse selection, causing high-quality goods to exit the market. This framework has been widely applied in food safety research to explain the structural vulnerabilities of food markets, where safety cannot be fully verified prior to consumption.

Food products are typically classified as credence goods, meaning that even after consumption, their safety cannot be fully verified by consumers. Darby and Karni (1973) theoretically showed that under conditions of information asymmetry, market equilibrium may involve fraud or underprovision of quality. This framework provides a robust theoretical basis for analyzing food safety issues in China, particularly in the context of major food safety incidents.

A representative study in the Chinese context is Pei et al. (2011), which analyzes the 2008 melamine-contaminated milk scandal. The study reveals that rapid market expansion in the dairy sector, fragmentation of the supply chain, and deficiencies in the regulatory system jointly contributed to the occurrence of a severe food safety crisis. This analysis reframes food safety incidents not merely as cases of corporate misconduct, but as economic phenomena arising from institutional incentives and regulatory failures.

Furthermore, empirical studies focusing on consumer behavior have demonstrated that demand for food safety in China is closely linked to market mechanisms. Ortega et al. (2011) show that Chinese consumers exhibit a significant willingness to pay

(WTP) for food safety attributes, suggesting that appropriate information provision and certification systems can generate market-based incentives for improving safety. Liu et al. (2013) further provide a comprehensive analysis of Chinese consumers' attitudes and behaviors toward food safety, highlighting that risk perception and trust in institutions play critical roles in shaping purchasing decisions. These findings indicate that not merely the quantity of information, but "trusted information," is crucial for the effective functioning of markets.

However, economic approaches also have several theoretical limitations. First, conventional models tend to assume rational actors and do not sufficiently incorporate socio-psychological factors such as risk perception, distrust, and the influence of media reporting. Second, the impact of macro-level trust structures—such as trust in institutions and government—on market transactions is not adequately captured by standard market failure theories. Third, in the context of the digital transformation of information flows, the role of online reviews and word-of-mouth in reshaping information asymmetry remains insufficiently theorized.

In summary, while economic studies on food safety in China have successfully identified institutional deficiencies and distorted incentive structures through the lens of information asymmetry and market failure, future research needs to move beyond market mechanisms alone. It is necessary to develop an integrated analytical framework that incorporates institutional trust, socio-psychological factors, and the digital information environment. In this sense, the re-theorization of food safety issues as part of a broader "economics of trust" constitutes a key direction for future research.

2.3.2 Sociological Approaches: Risk Society Theory and Trust Theory

Sociological studies on food safety issues in China have primarily developed within the frameworks of risk society theory and trust theory. Whereas economic approaches focus on information asymmetry and institutional incentives, sociological approaches are characterized by their emphasis on food safety as a form of social risk generated as a consequence of modernization, as well as on the transformation of trust structures.

1. **The Perspective of Risk Society Theory**

The theoretical foundation of risk society theory lies in Ulrich Beck's (1992) concept of the "risk society." Beck argued that while modernization generates material prosperity, it also diffuses *manufactured risks*—that is, risks that are artificially and institutionally produced—throughout society. Food safety issues are a typical example of such manufactured risks and are closely linked to the processes of industrialization, urbanization, and marketization.

In China as well, rapid economic growth and market-oriented reforms have accelerated the intensification of agriculture, the lengthening of food supply chains, and the sophistication of food processing technologies. These developments have created the structural conditions for risks that are difficult to manage. In particular, since the 2008 melamine-contaminated milk scandal, food safety has increasingly been understood not merely as a series of isolated incidents, but as a symbol of *structural risk* (Yan, 2012). In this context, food safety problems are interpreted not simply as issues of institutional immaturity or inadequate supervision, but as side effects of modernization itself.

Anthony Giddens (1990), meanwhile, argues that people in late modern society rely on trust in *abstract systems* in order to sustain everyday life. In food markets as well, consumers make purchasing decisions not by trusting individual producers directly, but by placing trust in abstract systems such as institutions, scientific expertise, and regulatory authorities. Food safety issues in China can therefore be analyzed as cases in which trust in such abstract systems has been destabilized.

2. **The Perspective of Trust Theory**

Within the framework of trust theory, food safety issues are understood not merely as problems of risk management, but as manifestations of a *crisis of social trust*. Niklas Luhmann (1979) defines trust as a mechanism for reducing social complexity, arguing that without trust, everyday transactions cannot be sustained. Since food is a good that is consumed routinely and repeatedly, the collapse of trust has implications that extend across the entire lifeworld.

In the Chinese context, trust is often regarded as a multi-layered structure. That is, trust exists in differentiated but interrelated forms, including:

- trust in firms,
- trust in places of origin and brands, and
- trust in government and institutions.

Han and Zhai (2022) analyze the effects of food safety issues on political trust and show that differences in trust toward local governments and the central government are linked to risk perception. This suggests that institutional trust is not singular, but hierarchically structured. Furthermore, in Chinese society, *guanxi* and social networks continue to play an important role. It has been suggested that when trust in formal institutions is weak, consumers tend to rely more heavily on information transmitted through family and personal networks. This provides a sociological perspective in which the weakening of institutional trust may lead to the strengthening of informal mechanisms of trust.

2.3.3 Policy Studies: Legal Systems and Administrative Reform

Policy studies on food safety issues in China have developed actively, particularly in response to major food safety incidents since the 2000s. In this line of research, food safety has increasingly been regarded as an important indicator of state governance capacity, and institutional reforms have been extensively analyzed.

1. **Development of the Legal System and Its Evaluation**

A major turning point in China's food safety legal framework was the enactment of the *Food Safety Law* in 2009. The 2015 revision further strengthened penalties and clarified liability to such an extent that it was described as "the strictest food safety law in history." In addition, the 2018 Party and state institutional reform led to a restructuring of the regulatory system. These institutional developments have been evaluated as attempts to integrate previously fragmented and overlapping regulatory powers and to move toward a more centralized management system (Liu, 2010).

Pei et al. (2011), in their analysis of the melamine-contaminated milk scandal, argued that the causes of the incident lay not simply in corporate misconduct, but in the fragmentation of supervisory authority, local protectionism, and insufficient information disclosure. Their work highlighted the need for comprehensive legal and institutional reform. Such studies theorize food safety problems as forms of *institutional failure* and emphasize the importance of improving legal and regulatory systems.

Moreover, the establishment of the State Administration for Market Regulation (SAMR) in 2018 aimed to integrate supervisory functions and improve administrative efficiency. This reform has been analyzed as an effort to overcome bureaucratic fragmentation by consolidating the oversight of food, pharmaceuticals, and quality inspection, thereby promoting more unified enforcement.

2. **Governance and Administrative Reform Perspectives**

A second major stream in policy research analyzes food safety from the perspective of governance theory, with particular attention to central–local relations and regulatory enforcement capacity. In its review of regulatory policy in China, the OECD (2021) emphasizes the importance of improving transparency, strengthening risk assessment capacity, and promoting evidence-based policymaking.

Similarly, the World Bank (2019), in its evaluation of food safety governance reform in China, reports that although institutional development has progressed, significant challenges remain at the local level, including insufficient enforcement capacity and a lack of supervisory resources.

Academic research has also pointed out that centralization does not necessarily lead directly to greater effectiveness. Han and Zhai (2022), for example, show that although institutional reforms have advanced, the restoration of public trust takes time. In other words, the improvement of legal systems and the reconstruction of social trust do not necessarily proceed in parallel.

3. **Theoretical Challenges**

Several theoretical challenges can be identified in policy studies.

First, many studies remain largely descriptive, focusing on institutional change and legal revision without adequately testing the causal effects of policy interventions. For example, systematic empirical research is still limited regarding the extent to which the 2015 revision of the Food Safety Law has actually contributed to reducing food incidents or restoring trust.

Second, the issue of institutional design—specifically, the balance between centralization and decentralization—has not yet been sufficiently theorized. Local governments face an incentive dilemma between promoting economic development and enforcing regulation, and this structural problem cannot be resolved merely through legal revision.

Third, policy studies tend to be biased toward the institutional supply side, that is, they are often government-centered. As a result, perspectives on multi-actor governance involving consumers, firms, and civil society remain relatively weak. Especially in the digital age, new forms of governance—such as online monitoring, information disclosure, and public oversight—have emerged, yet the theoretical frameworks to analyze them are still underdeveloped.

Overall, policy studies have made important contributions by systematically organizing China's food safety issues within the contexts of legal development and administrative reform, and by clarifying institutional progress such as the integration of supervisory systems, the strengthening of penalties, and the expansion of information disclosure. However, major theoretical challenges remain, including the gap between formal institutional development and substantive effectiveness, structural constraints arising from central–local relations, and the imperfect linkage between institutional reform and trust recovery. Future research should therefore move toward integrated policy analysis that includes the measurement of reform effects, the study of multi-level governance, and a stronger connection to consumer trust.

2.3.4 Limitations and Unexplored Areas in Existing Research on Food Safety Issues in China: The Need for a "Multi-layered Structure of Trust" Perspective

Existing research on food safety issues in China has developed along three major streams: economic studies (information asymmetry and market failure), sociological studies (risk society theory and trust theory), and policy studies (legal systems and administrative reform). However, these bodies of research still contain both theoretical and empirical limitations. To overcome these limitations, it is necessary to introduce the concept of the *multi-layered structure of trust* as a new analytical perspective.

1. **The Limitations of Institution-Centered Approaches**

Policy studies have generally evaluated positively the enactment of the *Food Safety Law* in 2009, its revision in 2015, and the integration of the regulatory system following the 2018 administrative restructuring (Liu, 2010; Pei et al., 2011). However, despite these institutional advances, they have not adequately explained why consumer distrust has persisted. In other words, a major theoretical issue remains: the formal strengthening of legal and regulatory systems does not necessarily coincide with the restoration of social trust.

2. **Theoretical Fragmentation and Insufficient Integration**

Economic studies, based on George A. Akerlof's (1970) theory of information asymmetry, have explained food safety problems as the outcomes of adverse selection and moral hazard. Since food is a typical *credence good*, consumers cannot fully verify its quality either before or after purchase, and thus markets cannot function efficiently without institutional supplementation (Darby & Karni, 1973).

By contrast, sociological studies, drawing on Ulrich Beck's (1992) theory of risk society and Niklas Luhmann's (1979) theory of trust, have interpreted food safety issues as side effects of modernization and as crises of trust. Anthony Giddens (1990) also points out that trust in abstract systems constitutes the foundation of late modern society, suggesting that food safety problems in China can be interpreted as manifestations of instability in abstract institutional trust.

However, these theoretical perspectives are often employed separately, and a framework capable of integrating market mechanisms, institutional design, and social trust in a coherent manner has not yet been sufficiently developed.

3. **The Reduction of Trust to a Single Layer**

In many existing studies, "trust" is measured as a single indicator. Yet, trust in Chinese society is in fact multi-layered. Han and Zhai (2022) show that food safety issues affect political trust in the central and local governments differently, highlighting the internal differentiation of institutional trust.

Moreover, when trust in formal institutions declines in China, dependence on informal networks—such as interpersonal ties and *guanxi*—tends to increase. This suggests that institutional trust, corporate trust, and interpersonal trust may stand in substitutive and complementary relationships. However, dynamic analyses of these interactions remain underdeveloped.

2.4 Theoretical Construction of a Three-Layer Trust Model for Food Safety

Existing research on food safety issues in China has accumulated important insights from multiple perspectives, including market institutions, policy reform, and consumer behavior. However, important challenges remain, particularly in that the analytical levels of the concept of trust have not always been clearly distinguished, and theoretical integration remains insufficient. Food safety problems are not merely issues of regulation; rather, they are social phenomena in which the *structure of trust* is formed and transformed across multiple levels. In light of these limitations, understanding food safety issues in China requires an analytical framework based on the *multi-layered structure of trust*. From this perspective, as illustrated in Fig. 2.6, this study proposes a three-layered trust model for food safety, consisting of psychological trust, institutional trust, and structural trust. These three layers do not form

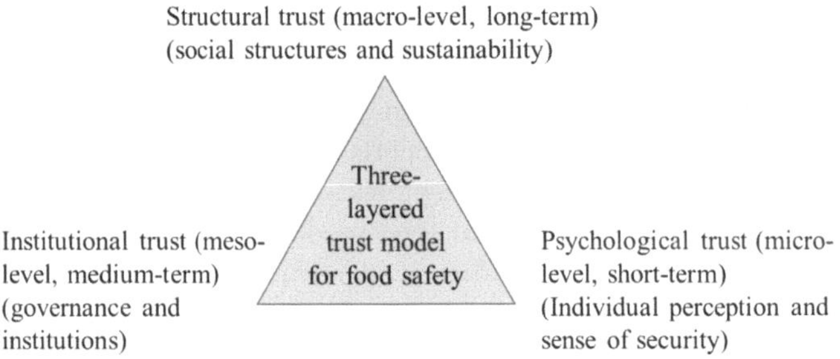

Fig. 2.7 A three-layer trust model in food safety (created by the author)

a simple lower–higher hierarchy. Rather, they constitute a *triangular equilibrium structure*, in which the collapse of any one point destabilizes the system as a whole.

The theoretical background of this three-layered trust model draws on three major perspectives: Anthony Giddens's (1990) theory of reflexive modernity, Ulrich Beck's (1992) theory of risk society, and Niklas Luhmann's (1979) theory of trust as a mechanism for reducing complexity. The present model can be understood as an integrative extension of these frameworks.

The distinctive feature of this model lies in the fact that it does not reduce trust to a mere psychological phenomenon. Instead, it redefines trust as a relational concept embedded in institutions and social structures (see Fig. 2.7).

2.4.1 Psychological Trust (Individual Level): Risk Perception Theory, Trust Formation Mechanisms, and the Need for Empirical Analysis

Within the analytical framework of this study, which conceptualizes food safety issues as a *multi-layered structure of trust*, the first layer is psychological trust at the individual level. This dimension refers to how consumers perceive food safety and how trust or distrust is formed through psychological processes.

1. **Risk Perception Theory**

As discussed in Sect. 2.1, risk perception theory provides the theoretical foundation for understanding psychological trust. Slovic (1987) demonstrated that risk is evaluated not only by objective hazard levels but also by subjective perception, highlighting that factors such as *dread* and *unknown risk* significantly influence risk judgments. In the context of food safety, even low-probability risks tend to evoke strong perceptions of fear, particularly when they are associated with health impacts or vulnerable populations such as children.

Furthermore, prospect theory developed by Kahneman and Tversky (1979) shows that individuals tend to overweight losses relative to gains. This provides a theoretical explanation for why food safety incidents, even when isolated, can exert disproportionate effects on consumer behavior. That is, a single negative event may lead to a generalized expansion of distrust due to cognitive biases.

Empirical evidence from China also supports this perspective. Ortega et al. (2011) demonstrate that consumers' risk perceptions significantly influence purchasing behavior and preferences for imported food products. This indicates that institutional improvements alone are insufficient; it is equally important to understand individual-level cognitive structures.

2. **Mechanisms of Trust Formation**

Psychological trust is not determined solely by the amount of information available; rather, it is shaped by multiple factors, including trust in information sources, past experiences, and social evaluations. Luhmann (1979) conceptualizes trust as a mechanism for reducing social complexity, enabling individuals to act under conditions of uncertainty.

Similarly, Giddens (1990) argues that trust in expert systems is essential in modern society. In the context of food safety, consumers form trust through abstract systems such as scientific expertise, regulatory authorities, and certification mechanisms.

From a psychological perspective, trust is often conceptualized as consisting of three key components: *competence*, *integrity*, and *benevolence* (Mayer et al., 1995). Consumers' evaluations of food firms and government institutions are likely structured along these dimensions.

3. **The Need for Empirical Analysis**

Although existing studies have provided important theoretical frameworks, empirical research that systematically measures and analyzes psychological trust in the Chinese food safety context remains limited. Much of the literature has focused on institutional evaluation and policy analysis, with insufficient attention to individual-level risk perception and trust formation processes.

In particular, several issues remain underexplored:

- the interaction between risk perception and institutional trust,
- the long-term effects of food safety incidents on trust formation,
- the influence of digital information (e.g., social media and online reviews) on psychological trust, and
- differences in perception across demographic groups such as age, education, and income.

Therefore, empirical studies using survey methods and experimental designs are needed. By measuring psychological trust, it becomes possible to examine how institutional reforms are reflected in individual cognitive structures, thereby providing a foundation for the three-layered trust model.

Psychological trust represents the starting point of the three-layered trust model in food safety. Risk perception theory explains the structure of consumers' subjective judgments, while trust theory conceptualizes decision-making mechanisms under uncertainty. However, in the context of food safety research in China, systematic empirical analysis at the individual level remains insufficient. Accordingly, this study aims to conceptualize psychological trust more clearly and to verify it empirically. Detailed analyses are presented in Chaps. 3 and 4.

2.4.2 Institutional Trust (Governance Level): Legal Systems, Regulatory Frameworks, Risk Communication, and the Significance of Comparative Institutional Analysis

The second layer in the three-layered trust model is institutional trust, which refers to trust in public governance structures, including governments, regulatory agencies, legal systems, and expert systems. Food safety is a typical domain of public risk, and market transactions can function stably only when institutional trust is established.

1. **Legal Systems and Regulatory Frameworks**

The foundation of institutional trust lies in clear and consistent legal systems, as well as their effective enforcement. In China, the enactment of the *Food Safety Law* in 2009, its revision in 2015, and the administrative restructuring in 2018 have promoted the integration of regulatory systems. These reforms aimed to replace the previously fragmented regulatory framework with a more centralized, risk-based governance system.

Theoretically, this corresponds to Giddens's (1990) notion of trust in *abstract systems*. Consumers do not rely on individual producers but rather on the reliability of institutional systems when making food choices. If institutional systems lack consistency, overall market trust is difficult to restore, even if individual firms are trustworthy.

Moreover, governance research suggests that institutional trust is closely related to regulatory transparency, accountability, and the fair enforcement of laws (OECD, 2021). Therefore, not only institutional design but also enforcement capacity and the allocation of supervisory resources are critical.

2. **Risk Communication**

Risk communication plays a central role in the formation of institutional trust. As Slovic (1987) shows, risk perception is strongly influenced by subjective factors; thus, the mere presentation of scientific data is insufficient to establish trust.

Beck (1992) further argues that in risk societies, asymmetries in knowledge between experts and citizens can generate distrust. Therefore, transparent information disclosure, interactive communication, and timely responses to incidents are essential for maintaining institutional trust.

In the Chinese context, cases have been documented in which delays in information disclosure and unclear assignment of responsibility following food safety incidents have led to declines in trust (Pei et al., 2011). This suggests that not only the existence of institutions but also their operation and communication practices are decisive for trust formation.

3. **The Significance of Comparative Institutional Analysis**

Comparative institutional analysis provides an important perspective for theorizing institutional trust. By comparing how food safety governance systems are designed and how trust is formed across different political and administrative systems, it becomes possible to clarify the institutional characteristics and challenges specific to China.

For example, the European Union adopts a separation principle between risk assessment bodies and risk management authorities, whereas China has strengthened an integrated administrative model. This difference should be examined in terms of the trade-off between the independence of expertise and administrative efficiency (World Bank, 2019).

Comparative institutional analysis is not merely about institutional imitation; rather, it serves as a methodological framework for understanding the interaction between institutional design and social trust. Since institutional trust is shaped by cultural and political contexts, it cannot be explained by a single universal model.

Institutional trust is thus formed through the development of legal systems and regulatory frameworks, transparent risk communication, and comparative institutional learning. Although reforms in China's food safety governance demonstrate significant institutional progress, a gap remains between formal institutional development and the restoration of social trust.

Accordingly, within the three-layered trust model, it is necessary to empirically examine the structure and dynamics of institutional trust, while taking into account its interaction with psychological trust.

2.4.3 *Structural and Sustainable Trust (Societal Level)*

The third layer of the three-layered trust model—*structural and sustainable trust*—refers to the long-term foundation of trust formed at the level of socio-economic structures, information environments, and normative systems. This layer encompasses and supports both psychological trust at the individual level and institutional trust at the governance level. In societies where food safety incidents occur repeatedly, trust cannot be stabilized through ad hoc institutional reforms alone; rather, the redesign of structural conditions becomes indispensable.

1. **Digital Economy and Platform Responsibility**

The advancement of the digital economy has fundamentally transformed both food distribution systems and the mechanisms of trust formation. In China, large-scale e-commerce platforms such as Alibaba Group and JD.com, as well as food delivery platforms such as Meituan, function as core infrastructures within the food supply system. These platforms go beyond simple transaction intermediation by performing functions such as vendor screening, quality disclosure, traceability assurance, and reputation systems based on consumer reviews. Through these mechanisms, they reduce information asymmetry and play a mediating role in trust formation.

Recent studies further suggest that such platforms increasingly function as *quasi-regulators*, complementing state regulation and assuming roles within private governance structures (Kenney & Zysman, 2016; Srnicek, 2017). Particularly in China, the rapid development of digital technologies and platform firms has interacted with state-led governance, giving rise to a hybrid governance structure characterized by the interplay of state, market, and platform actors (Cunningham et al., 2021). These structural transformations directly affect both institutional trust and consumers' risk perceptions in the domain of food safety.

While platform-based governance enhances trust through the aggregation and visualization of information, it simultaneously introduces new risks, such as algorithmic opacity and manipulation of evaluation systems (Kenney & Zysman, 2016). Therefore, under the digital economy, the institutionalization of platform responsibility becomes a critical determinant of sustainable structural trust.

2. **Information Disclosure and Transparency**

A fundamental pillar of structural trust is the institutionalization of transparency and continuous information disclosure. Beyond reactive accountability following incidents, it is essential to ensure the regular disclosure of inspection results, administrative sanctions, and corporate compliance performance. Luhmann (1979) conceptualizes trust as a mechanism for reducing social complexity. In environments where information is insufficient, consumers tend to overestimate uncertainty, leading to the expansion of distrust. Thus, institutionalized transparency functions as a structural condition underpinning both psychological and institutional trust.

Furthermore, Beck (1992) argues that in modern societies, trust in institutions is subject to reflexive scrutiny. In the context of food safety, the development of traceability systems and public information platforms contributes to the formation of long-term trust infrastructures.

3. **Linkages to the Sustainable Development Goals (SDGs)**

Structural and sustainable trust must also be aligned with the principles of sustainable development. The global framework for this alignment is provided by the Sustainable Development Goals (SDGs) adopted by the United Nations (UN, 2015). Food safety is closely related to SDG 3 (Good Health and Well-being), SDG 12 (Responsible Consumption and Production), and SDG 16 (Peace, Justice, and Strong Institutions). The SDGs should not be understood merely as policy slogans but as a theoretical axis for the integrated redesign of policies that combine risk management, environmental sustainability, and consumer protection (UN, 2015). Transforming food safety from a short-term issue of incident response into a long-term challenge of restructuring production, distribution, and consumption systems is essential for establishing structural trust.

Structural and sustainable trust is formed through three key elements:

1. the institutionalization of platform responsibility in response to the digital economy,
2. the normalization of transparency and information disclosure, and
3. the integrated redesign of policies based on the SDGs.

In other words, it is necessary to formally position the quasi-regulatory functions of platform enterprises, ensure the visibility and accessibility of information throughout distribution and transaction processes, and construct policy systems that integrate environmental, social, and economic dimensions of sustainability.

Such structural trust functions as a foundational condition that underpins both psychological trust—related to individual risk perception and behavior—and institutional trust—based on legal systems and governance structures. In this sense, food safety should not be understood merely as a matter of regulatory enforcement but must be redefined as a problem of designing sustainable social structures. This study thus extends food safety from a "regulatory issue" to a "design challenge of sustainable social systems."

As illustrated in Fig. 2.7, this model conceptualizes the three layers of trust not as a hierarchical structure but as an interdependent triangular structure (*triangular equilibrium structure*). Psychological, institutional, and structural trust are not independent elements but mutually reinforcing components forming the vertices of a triangle. Because of this interdependence, the collapse of any single dimension can propagate to the others, destabilizing the entire system. For example, a decline in institutional trust may undermine psychological trust and eventually lead to broader distrust in social structures.

Thus, the model explains how localized failures of trust can escalate into systemic distrust. By doing so, this study moves beyond a linear approach centered on regulatory strengthening and reconceptualizes food safety as a challenge of designing sustainable social systems. From the perspective of policy engineering, it provides a theoretical foundation for integrating institutional design, technological innovation, and social trust.

2.4.4 Spatial and Temporal Dimensions in the Three-Layer Trust Model

In addition to conceptualizing trust in food safety as a three-layered system—psychological, institutional, and structural—this model aims to provide an integrated analytical framework by incorporating two additional dimensions: the *spatial dimension* and the *temporal dimension*. Table 2.2 illustrates the correspondence between the three layers of trust and the spatial and temporal dimensions.

First, along the spatial dimension, trust can be classified into three levels: micro (individual), meso (institutional/organizational), and macro (societal structural). Psychological trust at the micro level refers to subjective trust based on consumers'

Table 2.2 Correspondence between the three-layered trust model and the spatial and temporal dimensions

Trust type	Spatial dimension	Temporal dimension
Psychological trust	Micro	Short-term
Institutional trust	Meso	Medium-term
Structural trust	Macro	Long-term

risk perception, experiences, and emotions, and represents the most direct and immediate response to food safety issues. Institutional trust at the meso level refers to trust in legal systems, regulatory oversight, corporate quality control, and certification systems, and is formed through social norms and rules. At the macro level, structural trust refers to long-term and foundational trust supported by the overall governance structure of society, the principles of sustainability, and the digital information environment.

In contrast, the temporal dimension represents the dynamic process through which trust is formed, accumulated, and stabilized over time. Psychological trust is formed in the short term and is highly volatile, as it is strongly influenced by incidents, media reports, and personal experiences. Institutional trust is formed in the medium term and is gradually accumulated through policy continuity and the stability of institutional operations. Structural trust, in turn, is formed in the long term and is sustained through the internalization of institutions and the transformation of social structures.

Thus, in this model, the three layers of trust—psychological, institutional, and structural—correspond not only to the spatial hierarchy of micro, meso, and macro levels, but also to the temporal processes of short-term, medium-term, and long-term development. Changes in psychological trust influence the evaluation of institutional trust, while the stability of institutional trust contributes to the formation of structural trust. At the same time, the strength of structural trust feeds back into both institutional arrangements and individual perceptions. In other words, psychological trust can be characterized as short-term and micro-level, institutional trust as medium-term and meso-level, and structural trust as long-term and macro-level. Accordingly, this model provides a theoretical framework that conceptualizes food safety issues not merely as problems of institutional design or risk management, but as processes of restructuring multi-layered trust through the integration of spatial hierarchy and temporal evolution.

2.5 Policy Implications: Redesigning Food Safety Governance Based on Trust Building

The theoretical analysis presented in this chapter demonstrates that food safety issues are not merely problems of technical risk management, but rather complex social phenomena shaped by the interaction between risk perception and trust structures.

Therefore, conventional regulation-centered approaches alone are insufficient, and an integrated policy design based on trust building is required.

First, from the perspective of psychological trust, policy responses must take into account the characteristics of consumer risk perception. Risk is evaluated not only in terms of objective probability but also through subjective factors such as dread and unknown risk. Therefore, ensuring scientific safety alone is not sufficient to alleviate consumer anxiety. It is essential to enhance the comprehensibility and emotional resonance of information through effective risk communication.

Second, from the perspective of institutional trust, it is important not only to establish legal frameworks but also to ensure transparency, consistency, and accountability in their implementation. Trust formation depends not on the mere existence of institutions, but on how they are operated and how information is disclosed. In particular, the timeliness of responses to incidents and the transparency of information disclosure are critical factors in restoring trust.

Third, from the perspective of structural and sustainable trust, it is necessary to redesign the socio-economic structures that underpin food safety. Specifically, this includes the institutionalization of governance functions of digital platforms, the normalization of traceability and information disclosure, and the construction of sustainable production and consumption systems based on the SDGs. These measures enable not only short-term recovery of trust but also the formation of a stable and long-term trust foundation.

Furthermore, the three-layered trust model proposed in this chapter provides important implications for policy design. The psychological, institutional, and structural layers are interdependent, and strengthening only one layer is insufficient to stabilize trust as a whole. Therefore, policies should not be confined to a single domain but must be designed by integrating the three dimensions of individual cognition, institutional design, and social structure.

In conclusion, food safety policy should be redesigned based on the complementary relationship between a regulation-centered approach and a trust-building approach. That is, while ensuring objective safety through regulation, it is necessary to integrate responses to risk perception and the cultivation of trust in order to achieve consumer reassurance. The theoretical framework presented in this chapter provides a foundation for such a policy transformation.

References

Akerlof, G. A. (1970). The market for "lemons": Quality uncertainty and the market mechanism. *The Quarterly Journal of Economics*, 84(3), 488–500. https://doi.org/10.2307/1879431

Barber, B. (1983). *The logic and limits of trust*. Rutgers University Press.

Bauer, R. A. (1960). Consumer behavior as risk taking. In R. S. Hancock (Ed.), *Dynamic marketing for a changing world* (pp. 389–398). American Marketing Association.

Beck, U. (1992). *Risk society: Towards a new modernity*. Sage.

Codex Alimentarius Commission. (2006). *Food safety risk analysis: A guide for national food safety authorities*. FAO/WHO.

Cunningham, S. M. (1967). The major dimensions of perceived risk. In D. F. Cox (Ed.), *Risk taking and information handling in consumer behavior* (pp. 82–108). Harvard University Press.

Cunningham, E., Doherty, M., & Frayman, H. (2021). *China's digital economy: The emergence of platform governance*. Oxford University Press.

Darby, M. R., & Karni, E. (1973). Free competition and the optimal amount of fraud. *Journal of Law and Economics*, 16(1), 67–88. https://doi.org/10.1086/466756

de Jonge, J., Frewer, L. J., van Trijp, H. C. M., Renes, R. J., de Wit, W., & Timmers, J. (2004). Monitoring consumer confidence in food safety: An exploratory study. *British Food Journal*, 106(10/11), 837–849. https://doi.org/10.1108/00070700410561423

de Jonge, J., van Trijp, H. C. M., Renes, R. J., & Frewer, L. J. (2008). Understanding consumer confidence in the safety of food: Its two-dimensional structure and determinants. *Risk Analysis*, 28(3), 729–740. https://doi.org/10.1111/j.1539-6924.2008.01042.x

FAO, & WHO. (1995). *Application of risk analysis to food standards issues: Report of the Joint FAO/WHO Expert Consultation*. World Health Organization.

Fischhoff, B., Slovic, P., Lichtenstein, S., Read, S., & Combs, B. (1978). How safe is safe enough? A psychometric study of attitudes towards technological risks and benefits. *Policy Sciences*, 9(2), 127–152. https://doi.org/10.1007/BF00143739

Frewer, L. J., Howard, C., & Shepherd, R. (1995). Genetic engineering and food: What determines consumer acceptance? *British Food Journal*, 97(8), 31–36.

Giddens, A. (1979). *Central problems in social theory: Action, structure and contradiction in social analysis*. University of California Press.

Giddens, A. (1990). *The consequences of modernity*. Stanford University Press.

Giddens, A. (1991). *Modernity and self-identity: Self and society in the late modern age*. Stanford University Press.

Han, G., & Zhai, Y. (2022). Risk society and the politics of food safety problems in China. *Japanese Journal of Political Science*, 23(2), 73–87. https://doi.org/10.1017/S1468109921000372

Kaplan, S., & Garrick, B. J. (1981). On the quantitative definition of risk. *Risk Analysis*, 1(1), 11–27. https://doi.org/10.1111/j.1539-6924.1981.tb01350.x

Kahneman, D., & Tversky, A. (1979). Prospect theory: An analysis of decision under risk. *Econometrica*, 47(2), 263–291. https://doi.org/10.2307/1914185

Kenney, M., & Zysman, J. (2016). The rise of the platform economy. *Issues in Science and Technology, 32*(3), 61–69.

Knight, F. H. (1921). *Risk, uncertainty and profit*. Houghton Mifflin.

Lichtenstein, S., Slovic, P., Fischhoff, B., Layman, M., & Combs, B. (1978). Judged frequency of lethal events. *Journal of Experimental Psychology: Human Learning and Memory*, 4(6), 551–578. https://doi.org/10.1037/0278-7393.4.6.55

Liu, P. (2010). Tracing and regulating China's food safety system. *Food Control*, 21(6), 770–775. https://doi.org/10.1016/j.foodcont.2009.11.005

Liu, R., Pieniak, Z., & Verbeke, W. (2013). Consumers' attitudes and behaviour towards safe food in China. *Food Control*, 33(1), 93–104. https://doi.org/10.1016/j.foodcont.2013.01.051

Lobb, A. E., Mazzocchi, M., & Traill, W. B. (2007). Modelling risk perception and trust in food safety information within the theory of planned behaviour. *Food Quality and Preference*, 18(2), 384–395. https://doi.org/10.1016/j.foodqual.2006.10.004

Luhmann, N. (1979). *Trust and power*. Chichester, UK: Wiley.

Mayer, R. C., Davis, J. H., & Schoorman, F. D. (1995). An integrative model of organizational trust. *Academy of Management Review*, 20(3), 709–734. https://doi.org/10.2307/258792

McAllister, D. J. (1995). Affect- and cognition-based trust. *Academy of Management Journal*, 38(1), 24–59. https://doi.org/10.2307/256727

Mitchell, V.-W. (1999). Consumer perceived risk. *European Journal of Marketing*, 33(1/2), 163–195. https://doi.org/10.1108/03090569910249229

National Bureau of Statistics of China. (2024, December 31). *Value added of China's agriculture and related industries takes up 15.34 pct of GDP in 2023*.

OECD. (2021). *OECD regulatory policy outlook 2021*. OECD Publishing. https://doi.org/10.1787/38b0fdb1-en

Ortega, D. L., Wang, H. H., Wu, L., & Olynk, N. J. (2011). Modeling heterogeneity in consumer preferences for select food safety attributes in China. *Food Policy*, 36(2), 318–324. https://doi.org/10.1016/j.foodpol.2010.11.030

Pei, X., Tandon, A., Alldrick, A., Giorgi, L., Huang, W., & Yang, R. (2011). The China melamine milk scandal and its implications for food safety regulation. *Food Policy*, 36(3), 412–420. https://doi.org/10.1016/j.foodpol.2011.03.008

Rousseau, D. M., Sitkin, S. B., Burt, R. S., & Camerer, C. (1998). Not so different after all: A cross-discipline view of trust. *Academy of Management Review*, 23(3), 393–404. https://doi.org/10.5465/amr.1998.926617

Slovic, P. (1987). Perception of risk. *Science*, 236(4799), 280–285. https://doi.org/10.1126/science.3563507

Srnicek, N. (2017). *Platform capitalism*. Polity Press.

Tversky, A., & Kahneman, D. (1974). Judgment under uncertainty: Heuristics and biases. *Science, 185*(4157), 1124–1131. https://doi.org/10.1126/science.185.4157.1124

United Nations. (2015). *Transforming our world: The 2030 agenda for sustainable development*. United Nations.

Wang, J., Mao, Y., & Gale, F. (2008). Chinese consumer demand for food safety attributes in milk products. *Food Policy*, 33(1), 27–36. https://doi.org/10.1016/j.foodpol.2007.05.006

Yan, Y. (2012). Food safety and social risk in contemporary China. *The Journal of Asian Studies*, 71(3), 705–729. https://doi.org/10.1017/S0021911812000678

Yamagishi, T. (1998). *The structure of trust: The evolutionary game of mind and society*. University of Tokyo Press.

Chapter 3
Determinants of Consumer Trust in Food Safety: An Empirical Study Using Covariance Structure Analysis

Fengyang Wang, Xuepeng Qian, and Weisheng Zhou

3.1 Research Problem and Objectives

As mentioned above, in the modern food system, the increasing complexity of food production, processing, distribution, and consumption makes it difficult for stakeholders throughout the food chain, including consumers, to collect and explain information about the goods they handle, making it easier to cause anxiety about food safety. In particular, since 2000, various incidents and accidents related to food quality and safety have occurred frequently in China, and consumers' lack of information and scientific knowledge on food safety, as well as extensive media coverage, have exacerbated concerns about food safety (Xiao, 2011; Liu & Xu, 2024). The seriousness of these incidents and accidents is that when a small-scale food safety incident occurs in a certain area, the scope of impact is not limited to the area where it occurred, but always has a large nationwide impact far beyond the region. In addition, as mentioned above, the social and psychological impact of food safety problems is greater than the direct health impact because the health effects of food safety incidents and accidents are exaggerated, and some consumers do not have the ability to judge the situation or respond to misinformation or misleading information about food safety (Liu & Xu, 2024).

In this study, we utilize the psychological security of Chinese consumers regarding milk and dairy products as a case in point. The Sanlu melamine scandal refers to

F. Wang
Xi'an Peihua University, Xi'an, China

X. Qian
Graduate School of Global Environmental Studies, Sophia University, Tokyo, Japan
e-mail: qianxp@sophia.ac.jp

W. Zhou (✉)
College of Policy Science, Ritsumeikan University, Kyoto, Japan
e-mail: zhou@sps.ritsumei.ac.jp

W. Zhou (ed.), *Food Safety in China*, https://doi.org/10.1007/978-981-92-3741-8_3

the contamination of infant formula and dairy products with the toxic substance melamine by Sanlu, a major Chinese dairy manufacturer, in 2008. From September of that year onwards, news reports emerged that several infants who had consumed Sanlu-branded infant formula had developed kidney stones. This caused widespread panic across China, and Sanlu faced severe social criticism (it subsequently went bankrupt). This incident led to a significant decline in consumption of domestically produced infant formula and products made from domestic milk powder, resulting in a sharp increase in demand for imported infant formula. Particularly in the infant formula market, foreign brands saw a marked expansion in market share. The occurrence of this melamine-tainted infant formula scandal caused many consumers to avoid milk and infant formula, significantly impacting the consumption of milk and dairy products. According to a survey conducted by China Central Television (CCTV) among Beijing residents, it was reported that 70% of respondents stated they would not purchase Chinese-made powdered milk by Beijing Morning Post, 28 February 2011. This consumer anxiety manifested as avoidance of domestic dairy products and a preference for imported powdered milk following the Sanlu incident, exerting a long-term impact on China's dairy market (Pei et al., 2011; Tian, 2012).

In addition, in a previous survey study in Nanjing City, approximately 70% of the respondents indicated their intent to abstain from purchasing dairy products from Company Y in the future, citing the dairy safety incident (illegal use of additives) involving the major dairy manufacturer Y. Moreover, more than 80% of the respondents expressed their unwillingness to purchase domestic dairy products (22.5%), and a significant proportion (60.4%) indicated a substantial reduction in their consumption of domestic dairy products (Tian, 2012). Even after that, consumers' concerns about the safety of domestic milk and dairy products had not yet been dispelled. In August 2016, the China Dairy Industry Association released the "China Dairy Industry Quality Report 2016" with the cooperation of China's Ministry of Agriculture, Ministry of Industry and Information Technology, Ministry of Commerce, National Health and Family Planning Commission, General Administration of Quality Supervision, Inspection and Quarantine, and State Administration of Food and Drug Administration. This report published monitoring data for liquid milk sold in the Chinese market. According to news reports on this publication, the pass rate of Chinese dairy product sample inspections in 2015 reached 99.5%. Monitoring results from 2009 to 2015 showed that quality and safety risks of domestic dairy products were under complete control, and no prohibited additives—such as melamine or protein hydrolysate—were detected in raw milk (China Daily, 2016). This indicates that, while the objective safety of domestically produced dairy products has greatly improved, consumers' subjective concerns still persist.

In recent years, changes have been observed in Chinese consumers' perceptions of and purchasing behaviour towards dairy products. Whilst attitudes towards domestic and imported dairy products are becoming increasingly polarised, there has been a certain degree of progress in restoring trust in domestic dairy products and improving purchasing intentions (Xu et al., 2020; Wei & Wang, 2023). The main trends are outlined below.

According to existing research, Chinese consumers exhibit a certain willingness to pay for imported dairy products, and food safety, place of origin, and brand are important factors in their purchasing decisions (Xu et al., 2020). On the other hand, recent surveys have shown changes in the knowledge and consumption behaviors of Chinese residents regarding dairy products, with certain improvements observed in both dairy intake and perceptions of domestically produced dairy products (Wei & Wang, 2023).

These studies indicate that current purchasing intentions for domestically produced dairy products in China can be summarised as follows. Against a backdrop of diminished trust stemming from past safety issues (e.g., the Sanlu melamine scandal), a persistent segment of consumers favoured imported dairy products. However, following subsequent enhancements in quality control and improvements to the distribution system, trust in the quality of domestic dairy products is gradually being restored.

Regarding the safety of such foods, the relationship between the characteristics of risk perception among general consumers (where a gap exists between subjective and objective risk perceptions) and the methods of information dissemination by the mass media and how consumers receive it often leads to excessive consumer reactions, reputational damage, and other forms of undue anxiety or unnecessary confusion. This risks undermining trust in the food safety system and can be said to readily generate consumer unease.

Based on the theoretical research in Chap. 2, consumer confidence in the food safety system is a factor in consumer anxiety. In other words, it is no exaggeration to say that food safety and ensuring consumer confidence in the food safety system are urgent issues in China. Therefore, research on ensuring the reliability of food safety among consumers in China is very significant. However, in recent years, there have been many studies focusing on the legal system, administration, and business morals related to food safety in China, but on the other hand, there have been few empirical studies on the reliability of food safety through consumer awareness surveys (Wang et al., 2016, 2022).

Based on the theoretical framework described above, this chapter aims to clarify the current state of consumers' trust in food safety in China and its determinants through a consumer awareness survey conducted in China from the consumer's point of view. In this part, we will construct and verify the reliability of food safety evaluation model and examine the practical significance of risk communication that seeks to improve food safety reliability.

3.2 Formulation of Hypothesis, Model Setting, and Analysis Methods

As mentioned above, "trust" in general is composed of "expectation of competence" and "expectation of intention" as two components that lead to a positive expectation that the other person will not have a terrible outcome for him, despite the existence of social uncertainty. In this study, we refer to the definition of Nakashima (2012) and refer to food safety trust when consumers believe that the relevant parties in the food chain are taking appropriate intentions and actions. This trust covers the trust in all those involved in the realization of food safety, including businesses involved in the food system (individuals and organizations involved in production, processing, distribution, and sales), governments, scientists, and experts.

Previous studies have generally argued that "variables related to ability evaluation: technical knowledge and ability," "variables related to motivation: transparency (not concealing production processes or information), objectivity (not being incompatible with a particular point of view), honesty (not making false representations), and consistency (no illegal history of the company)" are the determinants of traditional trust (Yamagishi, 1999; Nakayachi & Cvetkovich, 2008). Additionally, in the field of risk perception, Earle and Cvetkovich (1995) use the SVS model (Similarity–Value–Trust Model), which posits that perceived value similarity fosters trust. This model explains that in the process by which an entity (such as an individual, company, or government) comes to trust another, the perceived similarity of values, norms, and goals shared between the parties increases the predictability of the other's behavior and the expectation of goodwill, thereby forming trust. According to this model, people tend to trust others when they perceive that the other's ideals and values are similar to their own. If consumers can empathize with how risk management institutions (such as governmental authorities) perceive an issue, what they prioritize, and how they approach problem-solving, this can lead to increased consumer trust. In addition, Nakayachi and Cvetkovich (2008) integrated the SVS model and the traditional trust model, and argued that the evaluation of the other party's ability and the similarity evaluation of core values lead to trust. Nakashima (2012) discusses "factors that affect consumers' confidence in safety" and states that whether or not a person can be confident that what they are trying to eat is safe depends on consumer judgment (risk perception), scientific administrative judgment (risk assessment and management), business behavior (risk control), and It was argued that it depends on the actions of each stakeholder regarding the consumer's ability to self-manage (risk aversion). De Jonge et al. (2004, 2008) uses the "Consumer Food Safety Trust Monitor" that includes "intrinsic awareness" and "external influence" to determine that consumer trust is mainly related to food safety: governments, businesses, food safety incidents and problems (damage experience), mass media reports, which argued that it was influenced by the perception of production technology and capacity. Sato et al. (2005) confirmed that consumers who actively collect and examine food safety through a multi-index model focusing on the acquisition of knowledge about food safety contribute to restoring confidence in domestic safety.

According to these previous studies, consumers' confidence in food safety is not only caused by external factors (access to information from the outside; trust in others), but also by consumers' self-management (knowledge; risk awareness; trust in themselves; experience of food damage). This study utilised findings from prior research identifying factors influencing the reliability of food safety. It posited that food safety trustworthiness is determined by the degree of "trust" placed in the following: food safety administration (government)—encompassing the administrative framework, legal system, and regulations, norms, and standards related to ensuring food safety; businesses (farmers, processors, and retailers); the mass media, scientists, and experts; and consumers themselves. Consequently, the following four hypotheses were proposed.

Hypothesis 1 (H1): If consumers can trust food businesses (farmers, and those involved in processing and sales), the safety of the food they produce, process and sell will be more reliable.

Hypothesis 2 (H2): If consumers can trust food safety information provided by the mass media, scientists, experts, etc., food safety will be more reliable.

Hypothesis 3 (H3): If consumers can trust the food safety administration, food safety will be more reliable.

Hypothesis 4 (H4): The more self-management consumers have regarding food safety, the more reliability consumers have in food safety.

In other words, the reliability of food safety was set as the objective variable, and the explanatory variables "trust in business operators," "trust in the mass media," "trust in experts (scientists)," "trust in food safety administration," and "self-management of consumers" were set as determinants of food safety reliability. Furthermore, in this paper, we have prepared observational variables for estimating the food safety reliability, business operators, information factors, administrative factors, and consumer voluntary management factors assumed in the model (Fig. 3.1) as latent variables. In addition, we will analyze the covariance structure of this model using questionnaire survey data and verify its validity. Furthermore, based on the results of four hypothesis tests, we will grasp the determinants of reliability of food safety in China and propose policy proposals to improve reliability of food safety to alleviate concerns about food safety. SPSS19.0 (Japanese) and AMOS19.0 (Japanese) were used for statistical analysis.

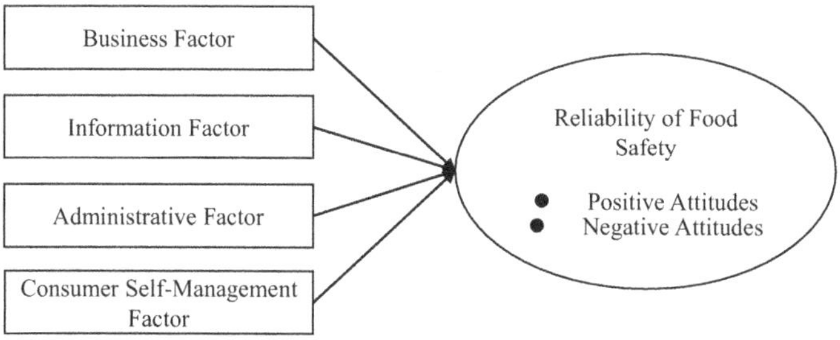

Fig. 3.1 Reliability of food safety analysis model

3.3 Questionnaire Survey and Results Analysis

3.3.1 Design and Implementation of Survey Items

In this study, the following survey items were designed based on the food safety trust questionnaire by De Jonge et al. (2008), considering the characteristics of China's population structure and food consumption. Each survey item consisted of (a) consumer personal attributes (8 items), (b) attitudes regarding the reliability of food safety (2 items: negative and positive), (c) consumer trust in the safety of various types of food (17 items: 17 types of foods), and (d) trust in each responsible entity for food safety (41 items) (see Table 3.1).

The questions consisted of multiple-choice answers and ranking answers. Consumers' personal attributes are single-choice responses, divided into gender, age, marital status, whether they have children, region, educational background, occupation, and annual income. Seventeen types of food were selected based on the food consumption habits and characteristics of Chinese consumers, and they were rated on a five-point scale from "strongly disagree" to "strongly agree."

The credibility of each management entity is based on the government (G), the farmer (F), the processing company (P), the seller (S), the mass media (M), the scientist and the expert (R), and the behavior of the consumer himself (C) was quantified on a five-point scale from "strongly disagree" to "strongly agree", and it was quantified on 1 ~ 5 points. In the analysis, "agree" and "strongly agree" were positive responses (positive attitudes), while "disagree" and "strongly disagree" were negative responses (negative attitudes).

To obtain the data necessary for the above analysis of the reliability of food safety, this study conducted research by establishing survey areas. Bearing in mind regional disparities within China (the gap between coastal and inland areas) and the urban–rural divide, Shanghai was selected as the survey area representing an economically and socially advanced region. Conversely, northern Anhui Province was selected as a case area representing an economically and socially less developed region (see Fig. 3.2).

In detail, the surveys were conducted in February and March 2015 in Shanghai (urban area) and northern Anhui Province (rural area), with the cooperation of local administrative departments and universities, targeting local residents, aged 18 or over, with 350 randomly selected men and women in each area (700 people in total). The number of valid responses was 482, and the valid response rate of this survey was about 69%. Respondent attributes and percentages are shown in Fig. 3.3.

Table 3.1 Mean structure of observed variables from the survey (N = 482)

Observed variable group 1 survey items	Mean structure		Observed variable group 2 survey items	Mean structure	
	mean	SD		mean	SD
G1: Government places importance on food safety	2.75	1.15	Trust in beef	2.91	0.94
G2: Government has the capacity to assess risks to food safety	2.59	1.14	Trust in pork	2.86	1.00
G3: Government has the capacity to monitor and control food safety	2.70	1.19	Trust in chicken meat	2.61	0.96
G4: Government has a complete law on food safety	2.81	1.05	Trust in fish meat	3.23	0.93
G5: Government should promptly release information on food safety incidents and accidents	2.47	1.13	Trust in meat products	2.45	0.91
G6: Trust food safety information published by governments	2.65	1.06	Trust in canned food	2.27	0.92
G7: Governments can ensure food safety	2.65	0.96	Trust in vegetables and fruits	3.52	0.89
F1: Farmers place importance on food safety	2.26	1.05	Trust in instant food	2.28	0.92
F2: Farmers have the knowledge to ensure food safety	2.65	0.96	Trust in eggs	3.42	0.95
F3: Farmers have production conditions that ensure food safety	2.48	1.02	Trust in sweets	3.04	0.90
F4: Farmers should disclose sufficient information on food safety	2.64	0.96	Trust in frozen food	2.40	0.88
F5: Trust food safety information published by farmers	2.42	1.00	Trust in milk and milk beverages	2.91	0.89
F6: Farmers can ensure food safety	2.49	1.01	Trust in bean products	3.26	0.93
			Trust in sugar	3.05	0.97
			Trust in tea leaves	3.48	0.93
			Trust in imported fresh food	3.17	1.04
			Trust in imported processed food	3.17	1.04

(continued)

Table 3.1 (continued)

Observed variable group 1 survey items	Mean structure		Observed variable group 2 survey items	Mean structure	
	mean	SD		mean	SD
P1: Processing companies place importance on food safety P2: Processing companies have the conditions and technologies to ensure food safety P3: Processing companies should fully disclose food information in food packaging P4: Trust the food information published by the processing company P5: Processing companies can standardize food additives P6: Processing companies can ensure food safety	2.62 2.37 2.77 2.40 2.26 2.16	1.01 1.08 0.92 0.88 0.92 0.87			
S1: Sellers place importance on food safety S2: Sellers can dispose of expired food as soon as possible S3: The seller shall sufficiently disclose information about the food to be sold S4: Trust the food information published by the seller S5: Sellers do not sell food products that have safety issues S6: Sellers can ensure food safety	2.21 2.21 2.22 2.27 2.25 2.02	0.93 0.92 0.86 0.86 0.85 0.89			

(continued)

Table 3.1 (continued)

Observed variable group 1 survey items	Mean structure		Observed variable group 2 survey items	Mean structure	
	mean	SD		mean	SD
C1: I place importance on food safety	2.21	1.05			
C2: Be aware of food safety	4.02	0.99			
C3: I have the knowledge and ability to ensure food safety	3.49	0.96			
C4: I can get food safety information	2.96	1.03			
C5: I can respond effectively if I experience a food safety incident	2.53	1.11			
C6: I can ensure food safety	2.71	1.06			
M1: The media places importance on food safety	2.58	1.03			
M2: The media can report food safety incidents and accidents as soon as possible	3.17	0.96			
M3: Food safety information from the media is well visible	3.10	0.98			
M4: Trust food safety information from the media	3.01	0.95			
M5: Media can ensure food safety	3.22	0.98			
R1: Experts place importance on food safety	2.68	1.07			
R2: Research on food safety by scientists and experts is accurate and valid	2.99	0.99			
R3: Food safety research results from experts are readily available	2.92	0.99			
R4: Trust food safety information and presentations from experts	2.50	1.00			
R5: Experts can ensure food safety	2.81	1.04			

3.3.2 Aggregate Analysis of Survey Results

According to the results of the questionnaire, 84.7% of the respondents had a negative response to attitudes regarding the reliability of food safety. The overall assessment of reliability of food safety tended to be relatively higher for negative responses (negative attitudes) than for positive responses.

Looking at the negative responses (negative attitudes) from the attributes of the survey subjects, the following specific characteristics can be mentioned. As shown in Fig. 3.4, males and females are almost indiscriminate. As shown in Fig. 3.5, people

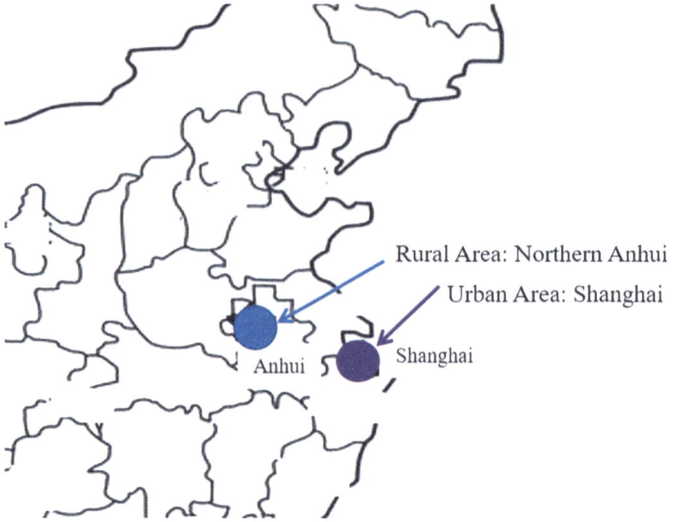

Fig. 3.2 Study areas

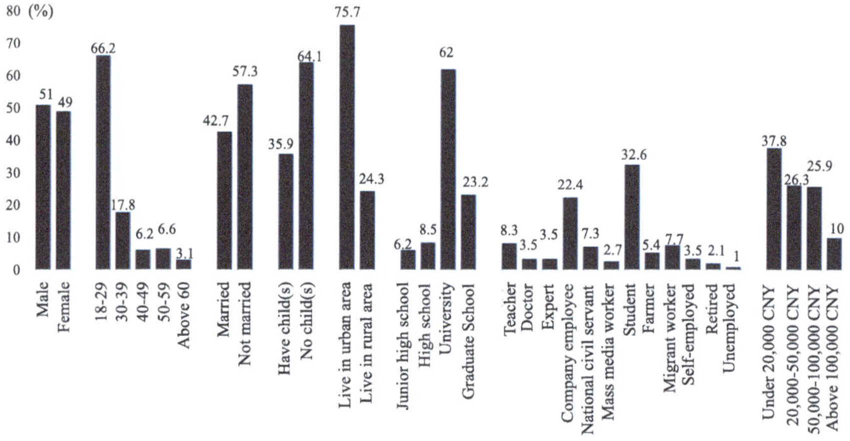

Fig. 3.3 Respondents' attributes (N = 482)

in their 30s have a relatively positive attitude toward food safety, while people in other age groups tend to have a negative attitude toward food safety as they get older. Or the distrust of the elderly is the strongest. In addition, married people are more distrustful of food safety than unmarried people (Fig. 3.6). People with children are more distrustful of food safety than those without children (Fig. 3.7). Compared to rural areas, the distrust of food safety in urban areas is stronger (Fig. 3.8). In terms of educational background, those who are below high school and secondary school are the least distrustful, and the higher the educational background, the more negative

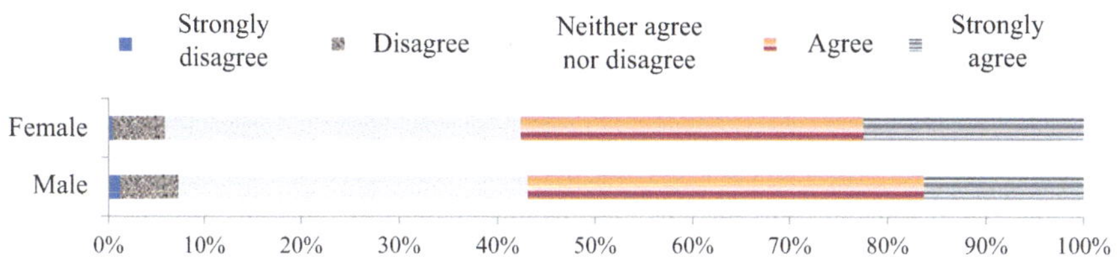

Fig. 3.4 Negative attitudes of reliability of food safety from the perspective of gender (N = 482)

attitudes they have (Fig. 3.9). In terms of annual income, those with disposable income of 100,000 yuan or more have a strong sense of distrust, while those with disposable income of 20,000 yuan or less have a relatively positive attitude (Fig. 3.10). Finally, as shown in Fig. 3.11, doctors, government employees, teachers, and retirees are the top four occupations.

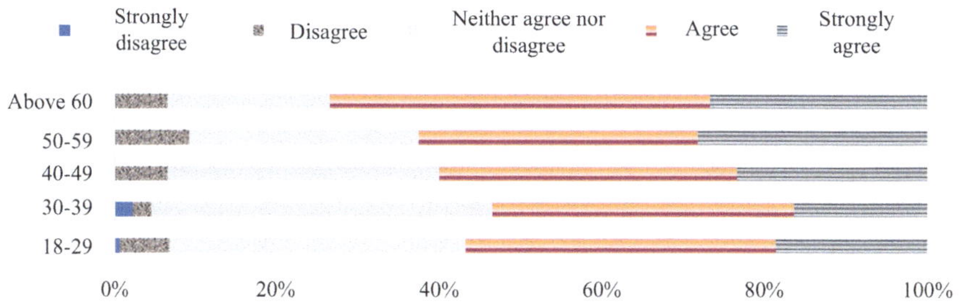

Fig. 3.5 Negative attitudes of reliability of food safety from the perspective of age (N = 482)

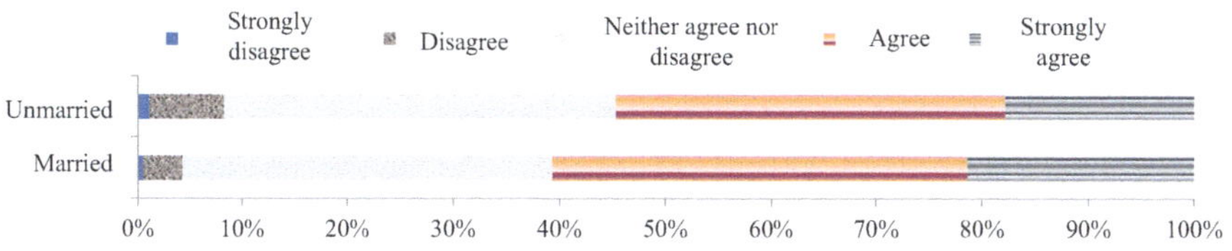

Fig. 3.6 Negative attitudes of reliability of food safety from the perspective of marital status (N = 482)

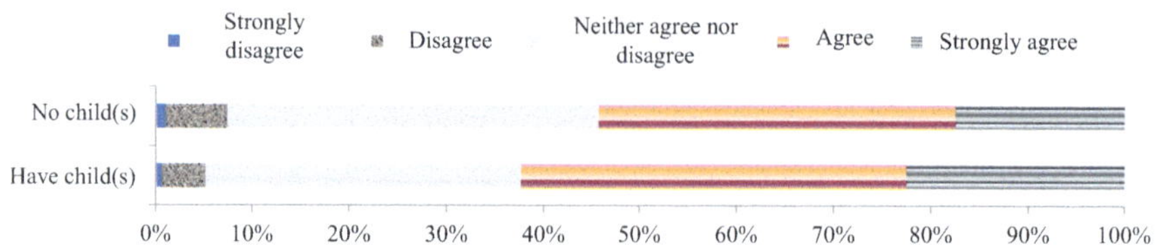

Fig. 3.7 Negative attitudes of reliability of food safety from the perspective of whether the respondents have child(s) (N = 482)

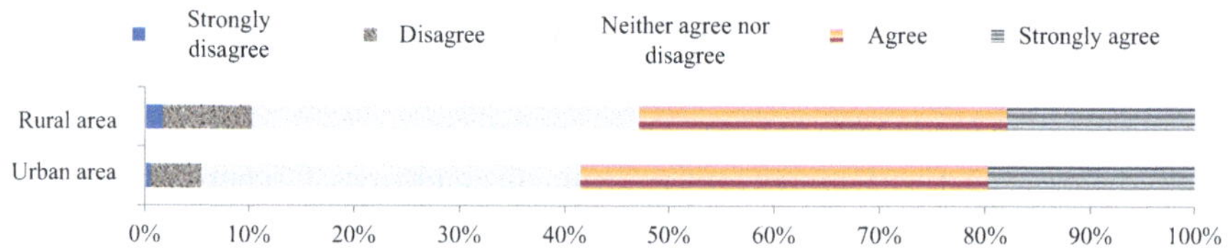

Fig. 3.8 Negative attitudes of reliability of food safety from the perspective of residential areas (N = 482)

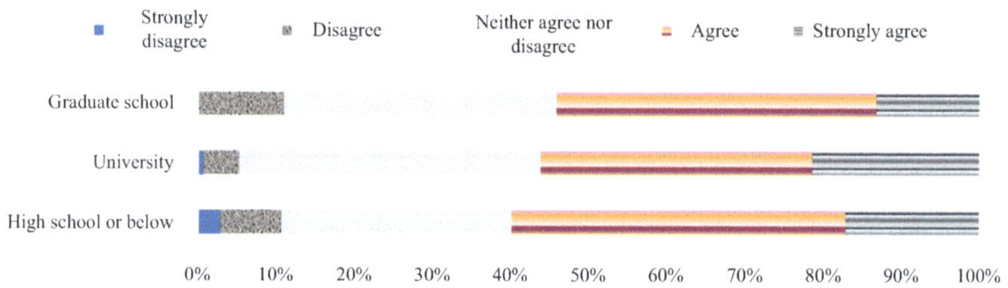

Fig. 3.9 Negative attitudes of reliability of food safety from the perspective of educational background (N = 482)

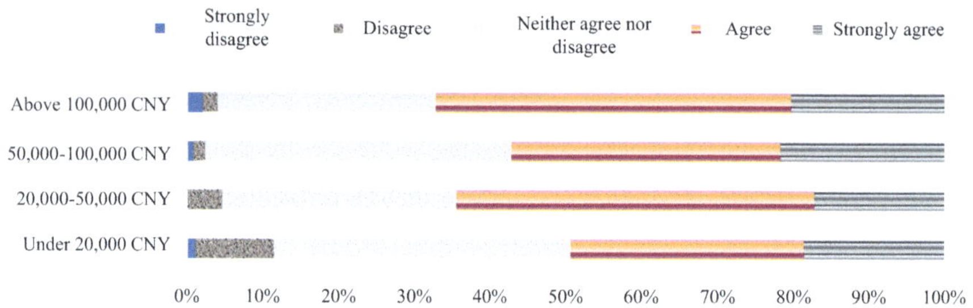

Fig. 3.10 Negative attitudes of reliability of food safety from the perspective of disposable income (N = 482)

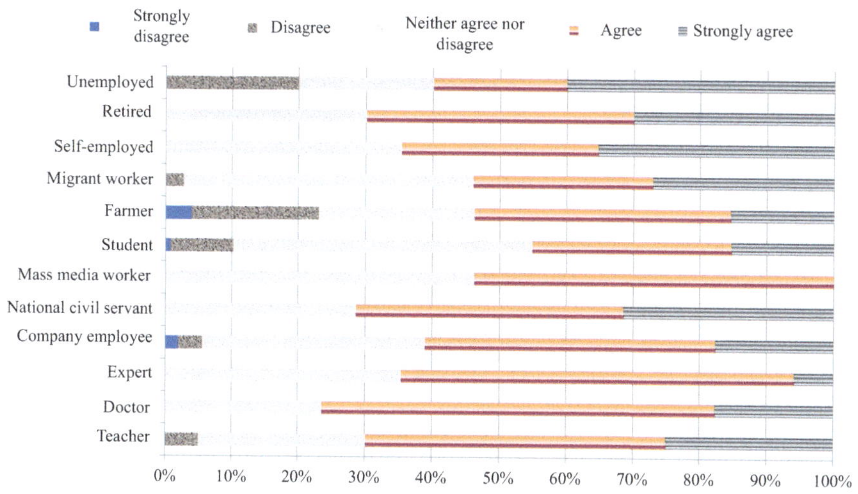

Fig. 3.11 Negative attitudes of reliability of food safety from the perspective of occupation (N = 482)

3.4 Analysis of Determinants of Reliability of Food Safety by Covariance Structure Analysis

3.4.1 Data Processing and Factor Analysis

Table 3.1 shows the mean and standard deviation of the observed variables. Furthermore, to explore the structure of trust in each stakeholder regarding food safety reliability, factor analysis was conducted on two groups of observation variables. In this factor analysis, principal component analysis is used, and only those items with an absolute value of 0.5 or more after varimax rotation are adopted.

First, we conduct a factor analysis on the reliability of each related entity. As shown in Tables 3.2 and 3.3, the Kaiser–Meyer–Olkin (KMO) and Bartlett tests were performed to determine whether factor analysis is effective. As a result, the KMO value is 0.883 ($\geq$0.6) and the α value of Cronbach is 0.946 ($\geq$0.7), which is more significant between the indicators than the Bartlett test ($x^2 = 12{,}880.719$, $p = 0.000 \leq 0.001$), so it is reasonable to perform factor analysis.

In terms of confidence in each entity related to food safety, four factors were extracted from 41 items. As shown in Table 3.4, the first factor was named "trust in businesses" because the factor loadings of the four variables "P4 (Trust the food information published by the processing company)", "P6 (Processing companies

Table 3.2 Reliability statistics (1)

Cronbach Alpha	Number of items
0.946	41

Table 3.3 KMO and Bartlett tests (1)

Kaiser–Meyer–Olkin measure of sample validity		0.883
Bartlett sphericity test	Approximate Chi-square	3833.044
	degree of freedom	136
	Significance probability	0.000

can ensure food safety)", "S4 (Trust the food information published by the seller)", and "S6 (Sellers can ensure food safety)" were large. In addition, the second factor was named "acquisition and reliability of external information" because the three variables "M1 (The media places importance on food safety)," "M4 (Trust food safety information from the media)," and "R2 (Research on food safety by scientists and experts is accurate and valid)" had high factor loadings (Wang et al., 2015). The third factor was named "perception and trust in monitoring and control capabilities" and consisted of three variables: "G1 (Government places importance on food safety)," "G3 (Government has the capacity to monitor and control food safety)," and "G6 (Trust food safety information published by governments)." The fourth factor was named "consumer self-management ability" and consisted of three variables: "C4 (I can get food safety information)," "C5 (I can respond effectively if I experience a food safety incident)," and "C6 (I can ensure food safety)". Regarding the details of the factor analysis, the first factor had an eigenvalue of 13.67 and a contribution rate of 33.33%; the second factor had an eigenvalue of 3.20 and a contribution rate of 7.82%; the third factor had an eigenvalue of 2.73 and a contribution rate of 6.65%; the fourth factor had an eigenvalue of 2.45 and a contribution rate of 5.98%, with a cumulative contribution rate of 53.77%.

Next, in the factor analysis on confidence in the safety of various foods, as shown in Tables 3.5 and 3.6, the KMO value is 0.927 ($\geq$0.6) and the α value of Cronbach is 0.896 ($\geq$0.7), and since there is a significant relationship between the indicators than the Bartlett test ($x^2 = 3833.044$, $p = 0.000 \leq 0.001$), it can be said that it is reasonable for factor analysis. As shown in Table 3.7, five factors were extracted from 17 items. The first factor was named "trust in fresh meat" because the three variables of "trust in beef," "trust in chicken," and "trust in pork" were high. The second factor was named "trust in processed foods" because the two variables of "trust in canned foods" and "trust in frozen foods" were loaded. The third factor, "trust in vegetables and fruits" and "trust in tea leaves," was named "trust in vegetables, fruits, and tea leaves." The fourth factor is named "trust in imported foods" for the two variables of "trust in imported fresh foods" and "trust in imported processed foods." The fifth factor was named "trust in dairy products" for the two items of "trust in milk and milk beverages" and "trust in soy products" (in China, milk, powdered milk, soy milk, and soy flour are classified as "dairy products"). For the details of the factor analysis, the first factor had an eigenvalue of 6.47 and a contribution rate of 38.03%; the second factor had an eigenvalue of 1.75 and a contribution rate of 10.29%; the third factor had an eigenvalue of 1.69 and a contribution rate of 9.96%; the fourth factor had an eigenvalue of 1.10 and a contribution rate of 6.46%; the fifth factor had

Table 3.4 Factor analysis of reliability for each entity related to food safety (N = 482)

	Trust in businesses	Acquisition and reliability of external information	Perception and trust in monitoring and control capabilities	Consumer self-management ability
G1: Government places importance on food safety	0.246	0.301	**0.648**	0.093
G3: Government has the capacity to monitor and control food safety	0.248	0.278	**0.664**	0.104
G6: Trust food safety information published by governments	0.253	0.270	**0.681**	0.128
P4: Trust the food information published by the processing company	**0.776**	0.158	0.141	0.070
P6: Processing companies can ensure food safety	**0.735**	0.151	0.196	0.109
S4: Trust the food information published by the seller	**0.768**	0.133	0.157	0.152
S6: Sellers can ensure food safety	**0.771**	0.052	0.194	0.240
C4: I can get food safety information	0.161	0.257	−0.009	**0.682**
C5: I can respond effectively if I experience a food safety incident	0.154	0.255	−0.018	**0.631**
C6: I can ensure food safety	0.096	0.312	−0.224	**0.600**
M1: The media places importance on food safety	0.164	**0.667**	0.151	0.048

(continued)

Table 3.4 (continued)

	Trust in businesses	Acquisition and reliability of external information	Perception and trust in monitoring and control capabilities	Consumer self-management ability
M4: Trust food safety information from the media	0.103	**0.648**	0.109	0.020
R2: Research on food safety by scientists and experts is accurate and valid	0.166	**0.666**	0.263	0.149

Note 1 The factor extraction method uses principal component analysis. The rotation method is a varimax method with Kaiser's normalization
Note 2 Only items with absolute factor loadings ≥ 0.5 after varimax rotation are shown

Table 3.5 Reliability statistics (2)

Cronbach Alpha	Number of items
0.896	17

Table 3.6 KMO and Bartlett Tests (2)

Kaiser–Meyer–Olkin measure of sample validity		0.927
Bartlett sphericity test	Approximate Chi-square	12,880.719
	degree of freedom	820
	Significance probability	0.000

an eigenvalue of 1.05 and a contribution rate of 4.42%, with a cumulative contribution rate of 69.19%.

Based on the factors with high loadings extracted from the factor analysis above, the four hypotheses can be concretely expressed as follows:

H1: If consumers can trust food businesses, they can also trust the safety of the food they produce and sell.
H2: The more food safety information consumers can rely on, the more trust in food safety.
H3: The more consumers perceive and trust the government's ability to monitor and control food, the more trust in food safety.
H4: The higher the awareness consumers have about food safety, the more trust in food safety.

Table 3.7 Factor analysis of trust of various foods for safety (N = 482)

	Trust in fresh meat	Trust in processed foods	Trust in vegetables, fruits and tea leaves	Trust in imported food	Trust in dairy products
Trust in beef	**0.729**	0.026	0.152	0.179	0.331
Trust in pork	**0.786**	0.240	0.296	−0.029	0.090
Trust in chicken meat	**0.783**	0.298	0.117	0.113	−0.019
Trust in canned food	0.223	**0.780**	0.085	0.037	0.077
Trust in frozen food	0.192	**0.629**	−0.028	0.186	0.389
Trust in vegetables and fruits	0.322	−0.029	**0.729**	0.074	0.140
Trust in tea (leaves)	0.046	0.133	**0.700**	0.267	0.168
Trust in milk and milk beverages	0.164	0.186	0.260	0.169	**0.759**
Trust in soy products	0.175	0.091	0.275	0.199	**0.755**
Trust in imported fresh food	0.133	0.131	0.147	**0.906**	0.160
Trust in imported processed foods	0.128	0.148	0.120	**0.902**	0.162

Note 1 The factor extraction method uses principal component analysis. The rotation method is a varimax method with Kaiser's normalization
Note 2 Only items with absolute factor loadings ≥ 0.5 after varimax rotation are shown

3.4.2 Empirical Models and Hypotheses Testing for Trust in Food Safety

Based on the results of the factor analysis, an empirical model was constructed (Fig. 3.12) in which the four factors of "trust in businesses," "acquisition and trust of external information," "perception and trust in monitoring and control capabilities," and "consumer self-management capability" were used as explanatory variables, with "reliability of food safety" as the ultimate objective variable. In the analysis, we removed non-significant paths and correlations to improve the goodness of fit, and adopted the model with the best fit as the estimation result. The evaluation results are shown in Table 3.8. All paths adopted were significant at the 5% level. Below is an overview of the empirical model using covariance structure analysis and its estimation results.

First, the evaluation results of the model in AMOS 19.0 are shown in Table 3.8. From the evaluation of the entire model, the $\chi2$ test is $\chi2$/df (degrees of freedom) = 3.01; GFI = 0.921, AGFI = 0.890, RMR = 0.051, and RMSEA = 0.065. In general, a GFI and CFI of 0.90 or higher and an RMSEA between 0.05 and 0.08 are considered acceptable; therefore, the fit indices of this model can be regarded as generally satisfactory (Schermelleh-Engel et al., 2003).

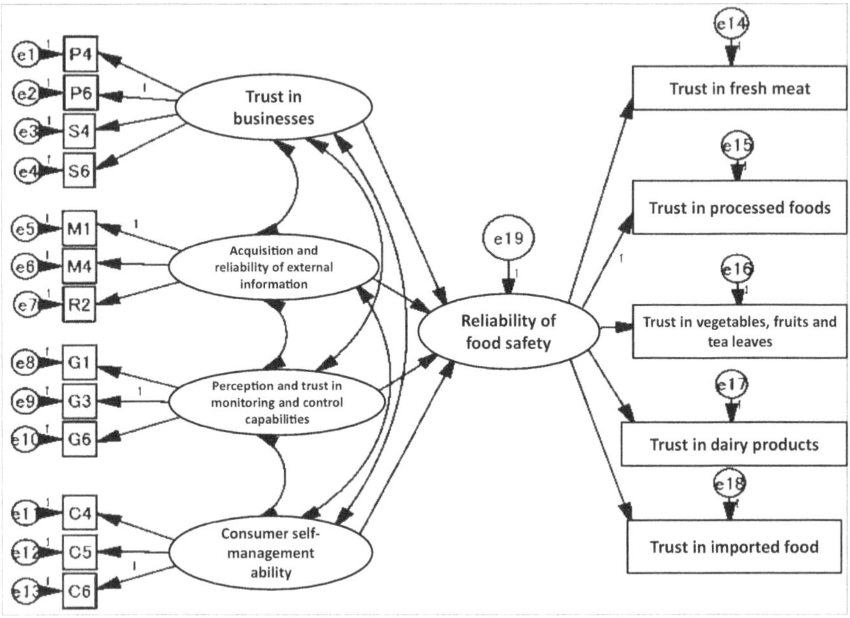

Fig. 3.12 Empirical model of trust in food safety (standardized)

Table 3.8 Overall model fit evaluation results

	RMR	GFI	AGFI	CFI	RMSEA
model	0.051	0.921	0.890	0.923	0.065
Saturation model	0.000	1.000	–	1.000	–
Independent model	0.263	0.414	0.352	0.000	0.211

In addition, the path coefficient values must be sufficiently large (in absolute terms) compared to zero. If it is close to zero, it indicates that there is no relationship between the two variables. A t-test is used to determine whether this coefficient is statistically significant. The results of testing the four modified hypotheses, based on the partial evaluation t-test for the empirical model ($|t| > 1.96$) and the p-value ($p < 0.05$), are shown in Table 3.9. Based on the correlation between "reliability of food safety" and the four explanatory variables, the aforementioned four hypotheses are veritifed. That is, H1 will be supported if there is a positive correlation with "trust in businesses". H2 will be supported if there is a positive correlation with "acquisition and reliability of external information", and H3 will be supported if there is a positive correlation with "perception and trust in monitoring and control capabilities". H4 will be supported if there is a positive correlation with "consumer self-management ability" related to food safety. As shown in Table 3.3.9, H1 and H2 were supported, but H3 was not supported ($t = 0.768$, $p = 0.443$), and H4 was

Table 3.9 Results of hypothesis testing

hypothesis	Standardized coefficients	correlation	t-test	P-value	Hypothesis testing results
H1	0.218	Positive correlation	2.920	0.004	Supported
H2	0.234	Positive correlation	2.778	0.005	Supported
H3	0.068	Positive correlation	0.768	0.443	Not supported
H4	−0.134	Negative correlation	−2.115	0.034	Not supported

Note If the absolute value of the t-value is greater than or equal to 1.96, it is significant at the 5% level ($p < 0.05$) in a two-sided test

not supported (standardized coefficient = –0.134, p = 0.034), indicating a negative correlation contrary to the hypothesis.

3.5 Discussion and Conclusion

This chapter provides an empirical analysis of trust in food safety and its determinants in China based on a consumer survey. In conclusion, the following points can be mentioned.

First, about 85% of the respondents had a negative attitude toward food safety reliability, which was relatively higher than a positive attitude. In terms of attributes, there is almost no difference between men and women. In terms of age, people in their 30s have a relatively positive attitude toward food safety, while people in other age groups have a stronger sense of distrust as they get older. Married people are more distrustful of food safety than unmarried people, and those with children are more distrustful of food safety than those without children. Compared to rural areas, there is a greater distrust of food safety in urban areas. In terms of educational background, people below high school and secondary school are the least distrustful, and the higher the educational background, the more negative attitudes they have. In terms of annual income, people with a disposable income of 100,000 CNY or more have a strong sense of distrust, while people with a disposable income of 20,000 CNY or less have a relatively positive attitude. In terms of occupational attributes, doctors, government employees, teachers, and retirees are the top four.

Furthermore, the following conclusions are drawn from the results of hypothesis testing of the determinants of reliability of food safety through an empirical model of food safety trust by covariance structure analysis.

First, it is extremely important to build a relationship of trust between businesses and consumers in response to consumers' distrust of food safety in China based on Hypothesis 1. In particular, in response to the fact that the lack of morals of some businesses may lead to mislabeling of food products and the use of illegal additives, it is necessary to take legal measures to correct inappropriate business activities and develop an administrative system in addition to food safety education. In this regard, two main policy directions can be proposed. First, food safety legislation and standards should be strengthened across the board, and penalties for illegal activities should be further strengthened. Second, there is a need to build a relationship of trust between food businesses and consumers by sharing information. There is an urgent need to institutionalize information disclosure through the introduction of a food traceability system in China.

Second, since Hypothesis 2 was confirmed and the standardized coefficient was the highest at 0.234 ($|t|=2.778$, $p = 0.005$), it can be seen that obtaining reliable information has the strongest positive impact on reliability of food safety. For consumers, the influence of acquired external (food safety) information and its reliability is considered to be very significant. Therefore, it is extremely important to develop methods for effective risk communication with consumers by the mass media and experts. Efforts should be made to train scientists and experts to support risk communication, to strengthen the exchange of information and opinions with consumers, to promote cooperation with the mass media, to present food-related information, and to establish forums for stakeholders to discuss basic issues related to food risks.

Third, Hypothesis 3 was not supported ($t = 0.768$, $p = 0.443$). In other words, the current situation is that food safety administration in China is not very effective in improving consumers' confidence in food safety, even if the government strengthens monitoring and control. In recent years, there has been steady progress in the comprehensive revision of the food safety law and the reform of the system in China, but there are still doubts about the effectiveness and reliability of the system. With the SVS (Salient Value Similarity) model of trust, risk management agencies need to make sure that consumers understand the management principles and methods to ensure food safety. Communication to deepen understanding from this perspective is considered to be the key to building trust.

Fourth, there is a weak negative correlation between consumer self-management ability and reliability of food safety (-0.134, $|t|=2.115$, $p = 0.034$), so Hypothesis 4 was not supported. In other words, the reliability of food safety will be reduced by increasing the ability of consumers to self-manage food safety. This indicates that knowledge about food safety and information gathering do not necessarily lead directly to increased trust and, in some cases, may even heighten caution or anxiety. Therefore, in consumer education, it is important not merely to increase the amount of information, but to focus on scientific understanding and appropriate risk communication (Sato et al., 2005; Wang et al., 2016). Conversely, it is believed that strengthening voluntary management of food safety, including the dissemination of prerequisite scientific knowledge and the careful collection of information, will suppress consumers' overreaction to food risks and affect their food purchasing behavior, so that they can purchase only reliable foods. This point will be reviewed in Chap. 4.

3.6 Policy Implications: Rebuilding from "Regulation-Centered" to "Trust-Centered" Approaches

The empirical analysis in this chapter revealed that consumer trust in food safety is not sufficiently secured solely by strengthening administrative supervision and institutional arrangements, as has been conventionally assumed; rather, it is determined primarily by trust in food businesses and the reliability of external information. Furthermore, an asymmetric structure was confirmed in which neither trust in government administration nor consumers' self-management capability directly translates into greater trust. These findings suggest the necessity of reorienting food safety policy from a "regulation-centered" approach to a "trust-centered" one.

First, because trust in food businesses significantly enhances consumer trust in food safety, policy should aim at "institutionalizing trust" through transparent and disciplined business conduct. That is, institutional assurance of corporate behavior—by upgrading traceability systems, reinforcing disclosure obligations, and imposing strict penalties for violations—must be combined with sustained information sharing that builds trust with consumers.

Second, the finding that acquisition and reliability of external information exert the strongest influence underscores the importance of information governance in food safety. It is essential for the mass media, scientists/experts, and administrative agencies to collaborate in providing accurate, easy-to-understand, evidence-based information and to strengthen two-way risk communication with consumers. Developing expert personnel, standardizing information dissemination, and creating forums for interactive dialogue constitute the foundation for trust-building.

Third, the lack of a significant effect of trust in the government's monitoring and control capabilities points to the limitations of the conventional competence-based model of trust. In line with the SVS model, an approach that shares policy philosophy and risk-perception frameworks with consumers, thereby building trust through value resonance, is required. Instead of simply tightening regulations, it is crucial to realize "understood governance" through enhanced transparency and accountability.

Fourth, the negative correlation between consumers' self-management capability and trust indicates that increased knowledge does not automatically lead to a greater sense of reassurance. Consumer education must therefore shift from the mere provision of information to the cultivation of adequate risk-judgment skills grounded in scientific understanding. Risk education that accounts for the uncertainty and relativity of risks, together with communication designs that mitigate overreactions, is called for.

In summary, food safety cannot be reduced to the responsibility of individual entities; it must be understood as an "institutional public good" that emerges from the interplay of institutions, information, and markets. Achieving this requires building cross-cutting governance that integrates regulatory policy, information policy, and education policy. In particular, leveraging digital technologies to visualize information and institutionalize trust will be a crucial policy challenge going forward.

Acknowledgements This chapter is an extensively revised, expanded, and updated English version of the author's article, "An Empirical Study on Consumer Trust in Food Safety in China and Its Determinants Based on a Consumer Awareness Survey," published in *Journal of Food System Research*, Vol. 22, No. 3, Japan Society for Food System Studies, 2015.

References

China Daily. (2016, August 19). Consumers still unsure of Chinese dairy quality. *China Daily*.

De Jonge, J., Frewer, L. J., van Trijp, H. C. M., Renes, R. J., de Wit, W., & Timmers, J. (2004). Monitoring consumer confidence in food safety: An exploratory study. *British Food Journal*, 106(10/11), 837–849. https://doi.org/10.1108/00070700410561423

De Jonge, J., van Trijp, H. C. M., Goddard, E., & Frewer, L. J. (2008). Consumer confidence in the safety of food in Canada and the Netherlands: The validation of a generic framework. *Food Quality and Preference*, 19(5), 439–451. https://doi.org/10.1016/j.foodqual.2008.01.002

Earle, T., & Cvetkovich, G. (1995). *Social trust: Toward a cosmopolitan society*. Praeger.

Liu, J., & Xu, W. (2024). An exploratory qualitative analysis of public awareness, communication dynamics, and behavioral patterns regarding food safety in Mainland China. *BMC Public Health*, 24, Article 2884. https://doi.org/10.1186/s12889-024-20382-3

Nakashima, Y. (2012). Institutional and economic systems for food safety and trust. *Journal of Food System Research*, 19(2), 55–61. https://doi.org/10.5874/jfsr.19.55.

Nakayachi, K., & Cvetkovich, G. (2008). Trust of risk managers: An integration of the SVS model and the traditional view of trust. *Japanese Journal of Social Psychology, 23*(3), 259–268.

Pei, X., Tandon, A., Alldrick, A., Giorgi, L., Huang, W., & Yang, R. (2011). The China melamine milk scandal and its implications for food safety regulation. *Food Policy*, 36(3), 412–420. https://doi.org/10.1016/j.foodpol.2011.03.008

Sato, K., Aizaki, H., Yoshikawa, H., & Sawada, M. (2005). *Shokuhin no anzen-sei ni kansuru chishiki ga shohisha no taido ni ataeru eikyo-BSE mondai no kyobunsan kozo bunseki* [The impact of food safety knowledge on consumer attitudes: Covariance structure analysis of the bse issue]. *Agricultural Information Research, 14*(1), 39–49.

Schermelleh-Engel, K., Moosbrugger, H., & Müller, H. (2003). Evaluating the fit of structural equation models: Tests of significance and descriptive goodness-of-fit measures. *Methods of psychological research online, 8*(2), 23–74.

Tian, M. (2012). *Nanjing municipal consumers' milk-quality-safety risk recognition and avoidance behavior* [Master's thesis, Nanjing Agricultural University].

Wang, F., Qian, X., & Zhou, W. (2015). An empirical study of consumers' trust in food safety in China: Focusing on factors influencing trust. *Journal of Food System Research, 22*(3), 299–304.

Wang, F., Zhou, W., Qian, X., & Nakagami, K. (2016). Risk communication aimed at improving trust in food safety in China: An empirical analysis based on a consumer survey on food safety information dissemination. *Social Systems Studies, 33*, 55–85. https://doi.org/10.34382/00003964

Wang, X., Xu, Y., Liang, M., Gao, J., Wang, J., Chen, S., & Cheng, J. (2022). Chinese Consumers' Trust in Food Safety Surveillance Sampling of Commonly Consumed Foods. *Foods, 11*(13), 1971. https://doi.org/10.3390/foods11131971

Wei, J., & Wang, J. (2023). Chinese residents' knowledge about and behavior towards dairy products: A cross-sectional study. *BMC Public Health*, 23, Article 374. https://doi.org/10.1186/s12889-023-15254-1

Xiao, P. (2011). China's milk scandals and its food risk assessment institutional framework. *European Journal of Risk Regulation*, 2(3), 397–406. https://doi.org/10.1017/S1867299X00001409

Xu, L., Yang, X., & Wu, L. (2020). Consumers' Willingness to Pay for Imported Milk: Based on Shanghai, China. International Journal of Environmental Research and Public Health, 17(1), 244. https://doi.org/10.3390/ijerph17010244

Yamagishi, T. (1999). *Anshin shakai kara shinrai shakai e [From a society of security to a society of trust]*. Chuo Koron Shinsha.

Chapter 4
Food Safety Information and Consumer Trust in China: An Empirical Analysis of Risk Communication

Fengyang Wang, Xuepeng Qian, and Weisheng Zhou

4.1 Awareness of the Problem and Research Objectives

As mentioned in the previous chapter, it is extremely important to build trust in the food safety system to alleviate consumer anxiety about food safety in China, especially information related to food safety from outside sources such as business operators, mass media, scientists, and experts (abbreviated as food safety information in the later context). The food safety information dealt with in this study is defined as the philosophy and purpose of food safety management, attitudes, regulations, standards, opinions, awareness, policy developments, explanations of the situation, business activities, safeguards, knowledge (latest research results), etc. regarding measures to protect consumers by each food-related entity (government, business operators, related associations, organizations, third parties, mass media, scientists, experts, doctors, health instructors, etc.) based on relevant laws, regulations, rules, and basic guidelines.

In other words, the strongest determinant of Chinese consumers' food safety reliability is the information factor about food safety for consumers. That is, it is necessary to improve the technical safety of the food safety system, and at the same time, to provide consumers with peace of mind and trust by clearly and accurately explaining the details of its management and measures, which is the expected role of "risk communication". Food safety risk communication is the exchange of information

F. Wang
Xian Peihua University, Xian, China
e-mail: fengyangwang@126.com

X. Qian
Graduate school of Global Environmental Studies, Sophia University, Tokyo, Japan
e-mail: qianxp@sophia.ac.jp

W. Zhou (✉)
College of Policy Science, Ritsumeikan University, Kyoto, Japan
e-mail: zhou@sps.ritsumei.ac.jp

W. Zhou (ed.), *Food Safety in China*, https://doi.org/10.1007/978-981-92-3741-8_4

and opinions from different standpoints among management entities (government, industry, consumers, academia, and other stakeholders) involved in risk assessment and risk management regarding the risks themselves, risk-related factors, and recognized risks in the entire process of risk analysis. It also includes a description of the results of the risk assessment and risk management decisions. However, risk communication in this study refers to the provision and sharing of food safety information from responsible persons and related parties to consumers. This information transmission basically includes three parties: the information sender (government, business, academia, etc.), the information recipient (consumer), and the information intermediary.

However, the current situation in China does not have the professional ability to analyze and evaluate the veracity and validity of the information available to many people regarding food safety on their own (Liu & Xu, 2024; Wang et al., 2016). Information asymmetry occurs when the information held by participants in a transaction or exchange is not equal, with one group having an "information advantage" and the other having an "information disadvantage," and the two groups are unable to share information (a situation in which there is uneven distribution of information) (Akerlof, 1970). Furthermore, consumers' intuitive judgment that ignoring "safety information" poses no threat to life or health leads them to pay great attention to information about "danger" and "anxiety," while information about "safety" and "security" is almost ignored. This kind of asymmetry and imbalance in food safety information and lack of consumer understanding often lead to psychological distrust among consumers. As mentioned in the previous chapter, this psychological distrust leads to a sense of anxiety and causes various social losses. Moreover, not only does the company that caused the problem undermine trust, but the anxiety caused by this can spread widely among consumers, leading to unfounded suspicions and unnecessary concerns to other food companies that are not originally relevant. In addition, excessive anxiety and unnecessary confusion such as overreaction by consumers and reputational damage often occur, and a situation where "bad money drives out good money" is likely to occur. For example, in the case of the avian influenza virus (H7N9) that occurred in China in March 2013, the WHO announced that the probability of people becoming infected with the avian influenza virus (H7N9) by eating chicken and chicken eggs is extremely low (World Health Organization, n.d.). In addition, the FAQ document on H7N9 jointly produced by the Food and Agriculture Organization of the United Nations (FAO) and the World Health Organization (WHO) explicitly states that "the virus is inactivated by cooking at normal temperatures (core temperature of 70°C or above), and cooked meat and eggs can be safely consumed" (Food and Agriculture Organization of the United Nations & World Health Organization, 2013). H However, various information about the H7N9 incident was circulated in the media and on the Internet, causing confusion and anxiety, and increasing consumer aversion to related foods. In other words, consumers lacked scientific knowledge and understanding of H7N9, resulting in a gap between subjective and objective risk perceptions. In addition, the overheated and erroneous coverage of the H7N9 incident further fueled consumer anxiety and fear. According to statistics from the China Livestock Industry Association, the incident caused a sharp decline in chicken and egg consumption in China, dealing a major blow to the poultry farming and sales industry nationwide. By the first half of 2013, the total losses of poultry businesses nationwide had exceeded 60 billion CNY (approximately 970 billion CNY), and it also had a negative impact on soybeans used for feed.

In particular, with the spread of the Internet, food safety issues have come to receive widespread attention due to the diversification of information means. According to the 37th Statistical Report on the Development of the Internet in China published by the China Internet Network Information Centre, as of the end of 2015, there were 688 million Internet users in China, with a penetration rate of 50.3%. Of these, the number of Internet users using mobile devices was 620 million, accounting for 90.1% of all Internet users. Table 4.1 presents a comparison of internet usage in China (2015 vs 2025) (CNNIC, 2016; CNNIC, 2025).

Due to the rapid spread of personal computers and smartphones, and the development of social media (SNS) such as Weibo and WeChat, the amount of information on the Internet, including food safety information, continues to increase explosively. Against this backdrop, in order to improve trust in food safety in China, we should first focus on food safety risk communication, which aims to build a relationship of trust with consumers. In particular, it is of great significance to fully clarify the status of the transmission of food safety information to consumers, which is closely related to food safety trust, and the issues of its reliability. However, at present, there are few empirical studies that deal with the correlation between consumer awareness and attitude toward food safety information and consumer food safety reliability from the perspective of food safety risk communication, and the ideal form of food safety risk communication. Therefore, in this study, we focused on the information factors of reliability of food safety, and grasped the status of consumer trust in food safety information (abbreviated as reliability of food safety information) through consumer awareness surveys in China targeting food safety information on the Internet, and constructed and verified a psychological factor evaluation model for reliability in food safety information. The objective of this study is to clarify the ideal form of risk communication.

Table 4.1 Comparison of internet usage in China (2015 vs 2025)

Indicator	2015	2025	Difference
Number of internet users	688 million people	1123 million people	+ 435 million people
Penetration rate	50.3%	79.7%	+ 29.4%
Number of mobile internet users	620 million people (90.1% of all internet users)	1116 million people (99.4% of all internet users)	+ 496 million people (+ 9.3%)

4.2 Analytical Framework

In addition to "risk assessment" and "risk management" based on scientific evidence, "risk communication" with the goal of improving understanding and reliability is also very important in the "risk analysis method" for improving food safety systems. As shown in Fig. 4.1, it is extremely important to provide accurate and scientifically based food safety information to the public through the media in risk communication regarding food safety policies and measures by stakeholders to ensure and publicize food safety. In particular, it is expected that the dissemination of food safety information by the government (government), food business operators, mass media (employees), scientists and experts, etc., will influence consumers' perception of food risks, deepen their understanding and trust in food safety and food safety policies and measures, generate trust, and relieve anxiety.

Regarding the characteristics of food safety information transmission in risk communication, previous studies suggest that general consumers judge the situation based on the reliability of the person to whom the information is disseminated, rather than the content of the food safety information. At the same time, it depends on the way the mass media reports it (Miyoshi, 2012). In other words, the system for disseminating information about food safety to consumers (source, content, method, etc.) is closely related to consumer trust. Regarding food safety, consumers are required to provide and share information based on reliable sources (information senders and information mediators). However, in China, food safety risk communication is still in the early stages of improving the reliability of food safety among consumers, and

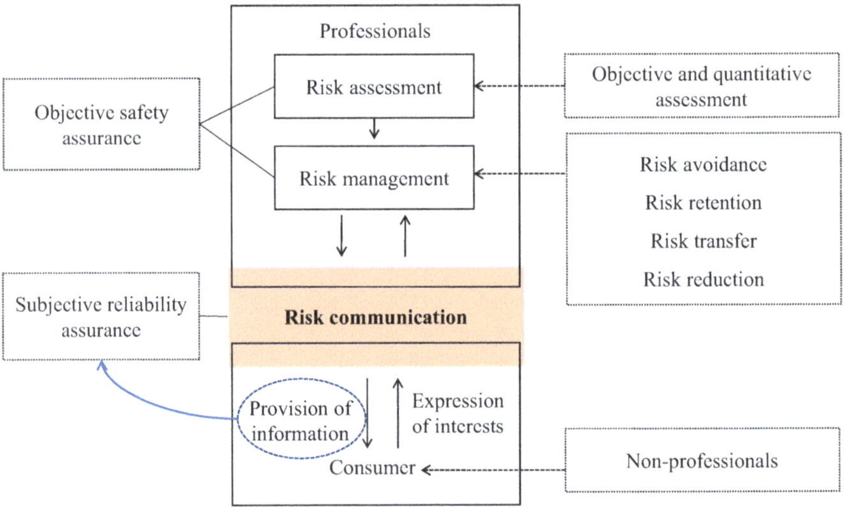

Fig. 4.1 Position of food safety information provision in risk communication (conceptual diagram). *Note* Risk communication here refers to the provision and sharing of food safety information to consumers by the person responsible and related parties

food safety management entities and related parties (government, industry, related associations, third parties, mass media, academia and experts, doctors, health guidance, etc.) are communicating and communicating food safety information. Issues such as the lack of promotion of understanding and the development of a sense of trust have been pointed out (Wang et al., 2022).

Therefore, in this study, we focus on information factors from the viewpoint of food safety trust, first examine the position of risk communication and food safety information provision, and then present the objectives of this study. Next, we will grasp the state of consumer trust in food safety information through a consumer awareness survey conducted in China. Then, we will conduct an empirical study on the influencing factors of food safety information reliability by covariance structure analysis using survey data. Finally, based on the results of the analysis, we will examine the ideal form of food safety risk communication aimed at improving the reliability of food safety in China.

4.3 Data Collection and Analysis Methods

4.3.1 Questionnaire Design and Implementation

In this study, we conducted a survey of consumer attitudes toward food safety information on the Internet in two representative regions of China (urban and rural) to understand consumer awareness of food safety information on the Internet. The subjects of the survey are interested in food safety information and can use the Internet (computers, smartphones, TVs, etc.). The survey items were designed with consideration for "personal attributes of consumers," "food safety information sources (information senders and information mediators)," "reliability of food safety information," and "risk perception and attitudes based on food safety information." The personal attributes of consumers, the main media of transmission of food safety information, and the route (site) of information dissemination were multiple choice, and personal attributes were divided into gender, age, marital status, family, region, educational background, occupation, and annual income, and the medium of transmission of information was divided into multiple choices: personal computers, smartphones, televisions, radios, newspapers, magazines, specialized books (including electronic versions), and others. In addition, the main channels of information transmission (information transmission websites) are also multiple selections, such as government homepages (GH), corporate homepages (CH), news portal sites (NH), sites of food-related industry groups, associations, and third-party organizations (AH), bulletin board and forums (BBS), Blog, Weibo and WeChat.

The survey on the reliability of food safety information and the risk perception and attitude of food safety information is a five-point Likert scale, and the confidence in food safety information is designed from "very distrustful" to "very trustworthy". The overall level of trust in online information was the subject of the survey, and the level of trust in each food safety information transmission website and the subject of information transmission (information disseminator) was also surveyed. In addition, we designed a scale for the reliability of food safety information and the tendency to respond to it based on previous research. Yamagishi and Komiyama's (1995) study also pointed out that there are two response tendencies related to trust: acquiescence tendency and extreme response tendency. In this study, the survey was conducted to investigate the reliability of food safety information and reaction trends by dividing food safety information from various parties when a food safety issue occurs into "high reliability (tendency to react positively)," "medium reliability (tendency to react negatively to some extent)," and "low reliability (tendency to react negatively)," using a five-point scale ranging from "Strongly disagree" to "Strongly agree." Specifically, "high reliability" means having a very high level of trust in food safety information and basically having no impact on food purchases. "Medium reliability" means that people have a sense of trust in food safety information, but have a certain amount of negative attitudes that influence food purchases. "Low reliability" means that even if they receive food safety information, they do not trust the relevant food safety or the policies and measures from the government, businesses, and other related parties, and not only do not purchase the food for a long time, but also tend to buy other substitutes.

To obtain the data necessary for analysis, the survey area was set up and the survey was conducted. In this study, after taking into account regional disparities (disparities between coastal and inland areas) and disparities between urban and rural areas in China, Shanghai was selected as the southeast coastal and urban areas (advanced areas for economic and social development) and the northern part of Anhui Province was selected as the inland and rural areas (less-advanced areas for economic and social development). According to the latest data from the time of the survey to the present, Shanghai's GDP for the 2024 fiscal year is approximately 5.39267 trillion yuan (an increase of approximately 5.0% year-on-year), with the proportion of the primary sector being extremely small. The city's per capita disposable income stands at 88,366 yuan, ranking first nationally (according to the Shanghai Municipal Bureau of Statistics and related published materials). Meanwhile, Anhui Province's 2023 GDP was approximately 470.506 billion yuan, showing year-on-year growth, while per capita disposable income in both urban and rural areas tends to be below the national average. The national average per capita disposable income for 2024 was 41,314 yuan, with urban areas at 54,188 yuan and rural areas at 23,119 yuan. Resident income in Anhui Province is slightly below the national average, with a persistent urban-rural disparity. The comparison between Shanghai City and Anhui Province is as follows.

From December 2015 to February 2016, a survey of 500 randomly selected males and females aged 18 and over living in the targeted study areas (1000 people in total) was conducted in Shanghai (urban areas) and northern Anhui Province (rural areas) in China with the cooperation of local administrative departments and universities. The number of respondents was 1000, and the response rate of the questionnaire was about 85.6% (the number of valid responses was 580, and the valid response rate excluding non-response and invalid responses was 58%), and the number of valid responses in rural areas was 81 and the number of valid responses in urban areas was 499. The response rate in rural areas was not low, but the number of valid responses was much lower than in urban areas. This may be determined by differences in income levels and educational levels due to regional differences.

4.3.2 Attributes of the Respondents

The attributes and proportions of respondents involved in this questionnaire survey are as follows. Characteristic features of respondents include a relatively low number of valid responses from rural areas and a high proportion of respondents with higher education qualifications. The reasons for this are: (1) The survey's prerequisite condition that respondents could access food safety information via the internet. According to CNNIC (2016) statistics, China's internet users are distributed regionally as follows: 26.9% in rural areas and 73.1% in urban areas. Regarding internet penetration by educational attainment, over 96% of those with a university or vocational college degree or higher are online; (2) In Shanghai City, one of the survey's areas, individuals with higher educational qualifications are relatively concentrated; (3) The number of valid responses from rural areas is lower than that from urban areas. Table 4.3 shows the attributes of the respondents of the survey.

4.3.3 Analytical Methods

In this study, we first analyzed questionnaire data to clarify the state of consumer trust in food safety information in China and its characteristics. In the analysis, a five-point scale was quantified as 1 ~ 5 points. "Very unreliable" / "I don't think so" and "Unreliable" / "I think it is different" are considered positive answers, and "Reliable" / "I think it is different" / "Very reliable" / " I don't think so" was considered a negative response. Next, in the psychological factor analysis, the reliability of food safety information is used as an endogenous latent variable, and the related observation variables are classified as " high reliability (positive response)," "medium reliability (somewhat negative response)," and "low reliability (negative response)" for food safety information and multiple observational variables related to the reliability of food safety information were also prepared (Table 4.4). Based on the results of the questionnaire survey data using principal component analysis, we constructed a psychological factor evaluation model for consumers' reliability in food safety information by covariance structure analysis, and verified its validity. Furthermore, we examined the relationship between the observed variables and latent variables of the model from the measurement results of the model, and analyzed the psychological factors of reliability of food safety information. In addition, we conducted a regression analysis of trust in food safety information and demographic variability, and examined the characteristics of reliability of food safety information by consumer attributes. Finally, based on the results of the analysis, we examined methods for communicating food safety information and how to communicate risks to improve consumers' confidence in food safety. SPSS19.0 (Japanese) and AMOS19.0 (Japanese) were used for analysis.

4.4 Analysis of Questionnaire Survey

4.4.1 Acquisition and Reliability of Food Safety Information

4.4.1.1 Transmission Media of Food Safety Information

In today's advanced information age, the means of information transmission have diversified and the scope has expanded, and the speed of transmission and the amount of information received have become faster and larger. From the survey results, the transmission medium for receiving food safety information of consumers in both regions of China is shown in Figure 4.2. The largest number of people who receive food safety information on their cell phones and smartphones is about 76.2%. Computers and televisions came in second and third place, respectively, with fewer newspapers, magazines, and books (including digital versions), and the least number of radio users. Furthermore, by region, both urban and rural areas have the largest number of cell phone and smartphone users, and the lowest number of radio users.

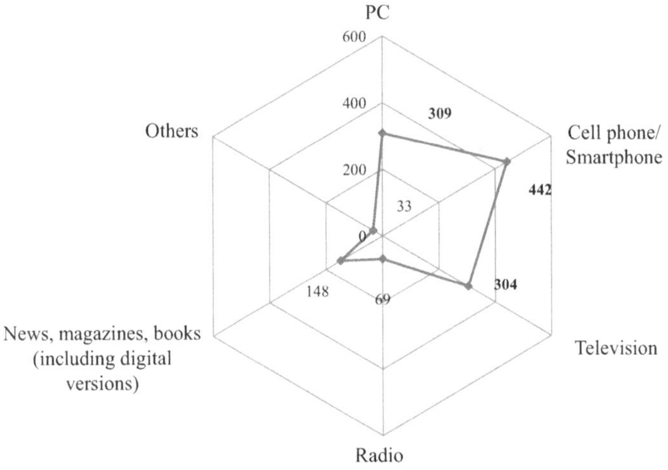

Fig. 4.2 Transmission medium for receiving food safety information from consumers (N = 580). *Note* This survey item is multiple choice. *Source* Created by the author based on survey results

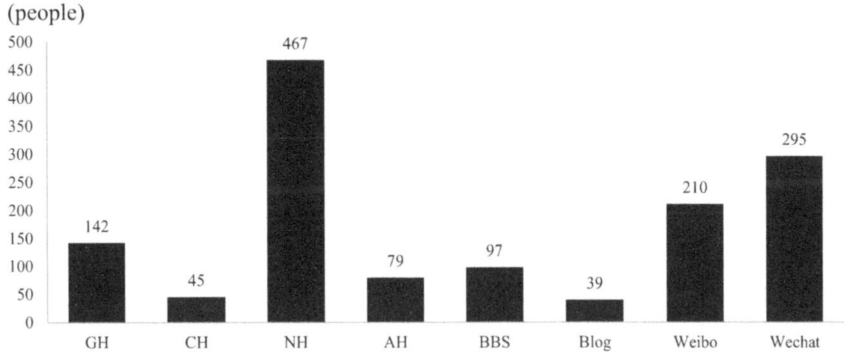

Fig. 4.3 Media for receiving food safety information for consumers (N = 580). *Note* This survey item is multiple choice. *Source* Created by the author based on survey results

4.4.1.2 Website for Disseminating Food Safety Information

With the spread of the Internet, especially the rapid increase in the number of smartphones, the Internet media that convey food safety information have also diversified. Figure 4.3 shows websites dealing with food safety information for consumers in both regions of China. According to the findings, news portal sites (NH) such as Sina (http://www.sina.com.cn), Fenghuang (http://www.ifeng.com), Tencent (http://www.qq.com), Net Ease (http://www.163.com) are the most important channel for obtaining food safety information, and the sophistication and development of Weibo, WeChat, bulletin boards and forums (BBS), SNS will replace traditional blogs as the main source of food safety information. There were not many people who used the government website (GH) as a source of information, ranking the fourth. There are few sources of information from corporate websites (CHs) and food-related industry groups, associations, and third-party organizations (AH). Furthermore, by age, the older the population, the more people receive food safety information from GH, while the fewer people use social networking sites such as WeChat and Weibo as a source of information (Fig. 4.4). According to the characteristics of each age group, it is considered necessary to provide effective information by combining appropriate information transmission websites.

4.4.1.3 Reliability of Food Safety Information

According to the survey results, about 27.2% of consumers in both regions of China had a "negative attitude" regarding their reliability in food safety information on the Internet, as shown in Fig. 4.5, and 10.2% of them chose "very unreliable" and 17.1% chose "unreliable." About 17.8% of respondents had a positive attitude, of which 3.3% chose "very reliable" and 14.5% chose "reliable." However, the remaining 55% of the respondents were ambiguous and chose "cannot say either way." If trust is to

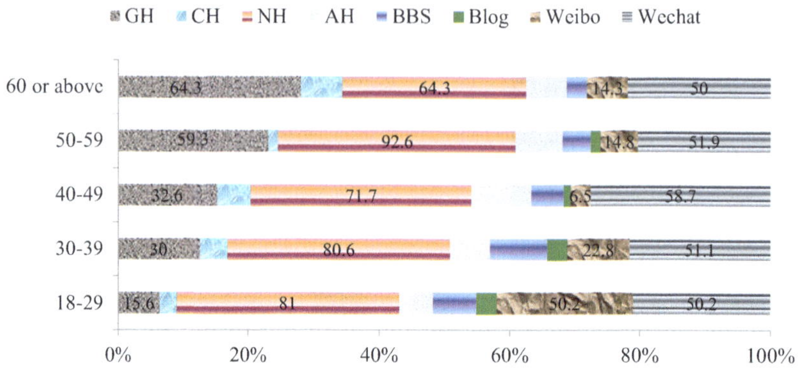

Fig. 4.4 Media sources of food safety information obtained by people of different age groups (N = 580). *Note* This survey item is multiple choice. *Source* Created by the author based on survey results

be gained, it is considered extremely important to stand more in the consumer's shoes and to transmit and communicate information that is "empathetic". As mentioned above, the theoretical basis for this is that consumers form emotional trust based on "empathy."

In addition, Figs. 4.6 and 4.7 show the reliability of the eight main media sources and the eight entities that transmit food safety information from the questionnaire results. Specifically, as shown in Fig. 4.6, consumers have the highest trust in food safety information (administrative information) obtained from GH, and relatively

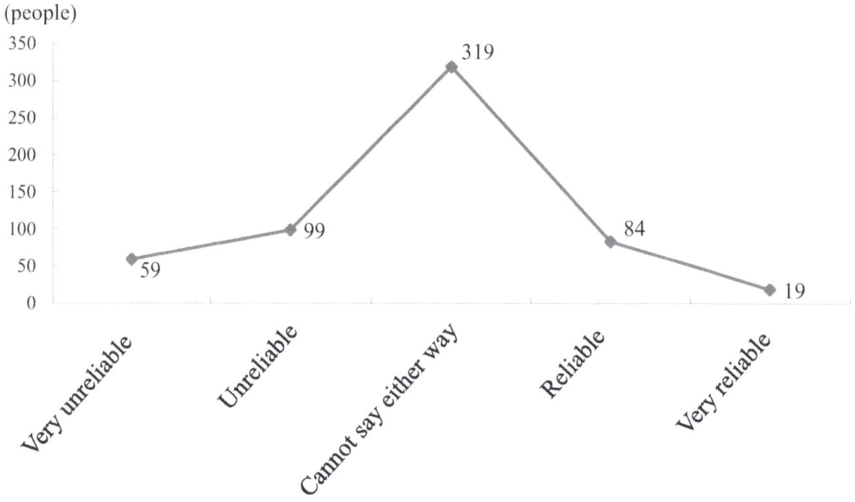

Fig. 4.5 Overall reliability in food safety information on the Internet (N = 580). *Source* Created by the author based on survey results

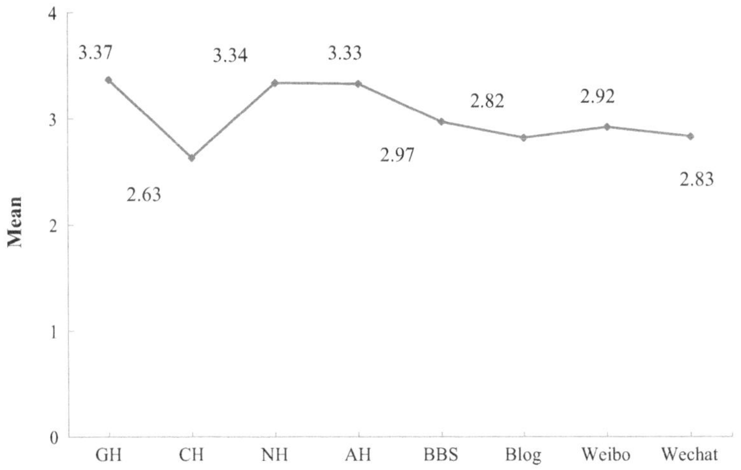

Fig. 4.6 Reliability in media sources that receive food safety information (N = 580). *Note* The reliability score is a mean value quantified on a five-point scale. *Source* Created by the author based on survey results

high trust in information from NH and AH. The reliability of the information from CH is the lowest. In other words, although the amount of information from social networking sites such as WeChat, Weibo, and BBS has increased rapidly, the improvement in reliability has not kept pace with the increase in the amount of information. On the other hand, when it comes to consumers' trust in the sources of food safety information, the trust in information that originates from doctors, the government, and close friends is high, and the trust in food businesses is the lowest (Fig. 4.7).

4.4.2 Trust in Food Safety Information and Trends in Responding to Food Safety Issues

From the survey results, consumers react differently to the food safety information they receive (Fig. 4.8). Therefore, the reliability of consumers' food safety information can be inferred from the different reactions. In designing the survey, we divided consumer reliability of food safety information from each management entity and related party in the event of a food safety problem into "high reliability (tendency to react positively)," "medium reliability (tendency to react negatively to some extent)," and "low reliability (tendency to react negatively)." In addition, the people belonging to the three reliability intervals were divided into "high reliability group," "medium reliability group," and "low reliability group." Based on the results, there were two levels of "medium reliability" (respondents who can understand the facts quickly, avoid related foods as much as possible, do not buy in a short period of time, and buy from places they trust) and "low reliability" (respondents answered that they do not

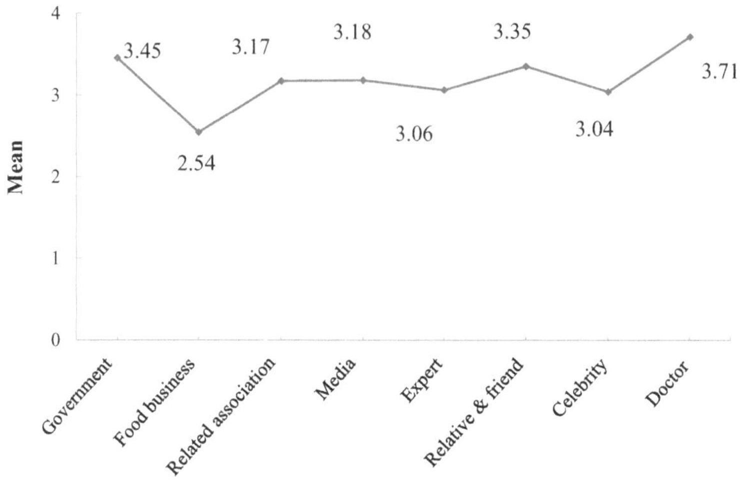

Fig. 4.7 Reliability in the sources of information (N = 580). *Note* The reliability of the sender of information is an average value quantified on a five-point scale. *Source* Created by the author based on survey results

purchase for a long period of time, and buy substitutes). However, there are few people who have "high reliability" (basically answered that there is no impact). Therefore, (1) consumers who have experienced food safety damage, or many consumers who perceive a food safety problem, tend to have low reliability in all food safety information regardless of the source, information medium, or content; (2) distrust in food safety information from business operators is naturally a problem for companies causing scandals, but once a problem occurs, the entire food industry is seriously affected. It leads to distrust in the safety of all related foods, not just those with safety problems, and the tendency to avoid related foods and buy substitutes is evident. This is due to (1) the fact that consumers do not know exactly what the problem is and have a vague sense of anxiety that makes them want to avoid purchasing not only the problematic product but also a wide range of adjacent products, and (2) the availability of substitutes makes it possible to avoid them.

From this result, it is possible to elucidate the case of the "melamine-contaminated powdered milk incident" described above. In other words, it is thought that it can explain the overreaction of consumers and reputational damage due to food safety issues such as the "melamine-laced milk powder" incident that occurred in China in 2008 mentioned in Chap. 1. The outbreak of the incident has caused many consumers to react sensitively and to avoid the consumption of milk and formula, which has had a significant impact on the consumption of domestic milk and dairy products.

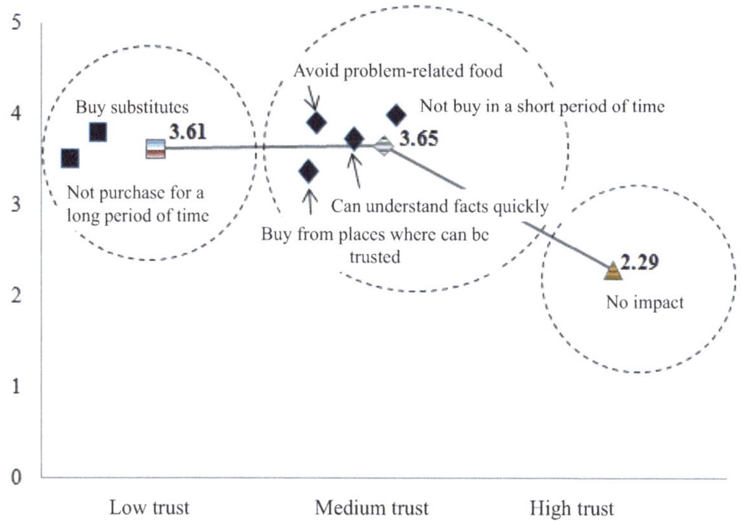

Fig. 4.8 Reliability and response trends to food safety information (N = 580). *Note 1* Based on the design of the survey, consumers' reliability in food safety information was divided into "high reliability (tendency to react positively)," "medium reliability (tendency to react negatively to some extent)," and "low reliability (tendency to react negatively)." *Note 2* The data is an average value quantified on a five-point scale from "Strongly disagree" to "Strongly agree." *Source* Created by the author based on survey results

4.4.3 Evaluation of Psychological Factors on the Reliability of Food Safety Information by Covariance Structure Analysis

4.4.3.1 Model Settings and Variable Definitions

In this study, the reliability of food safety information is considered as an endogenous latent variable, and the associated observational variables are Y_1: high reliability in food safety information, Y_2: medium reliability in food safety information, and Y_3: low reliability in food safety information. Other observation variables are shown in Table 4.2.

4.4.3.2 Model Fitting and Analysis

a. Examination of variable reliability and validity

First, as shown in Tables 4.5 and 4.6, in order to understand whether factor analysis is effective, the Kaiser-Meyer-Olkin (KMO) and Bartlett tests were performed, and the KMO value was 0.862 ($\geq$0.6), Cronbach's α value is 0.892 ($\geq$ 0.7) and the Bartlett

Table 4.2 GDP comparison between Shanghai City and Anhui Province (2014 vs 2024)

Area	Year	GDP (US dollar)	Per capita GDP (US dollar)	Notes
Shanghai City	2014	378.78 billion	18,700	Population: approximately 20.24 million Exchange rate: 1 CNY ≈ 0.1607 USD
	2024	751.2 billion	37,500	Population: approximately 20.03 million Exchange rate: 1 CNY ≈ 0.1393 USD
Anhui Province	2014	335.19 billion	5880	Population: approximately 57 million Exchange rate: 1 CNY ≈ 0.1607 USD
	2024	704.9 billion	11,700	Population: approximately 60 million Exchange rate: 1 CNY ≈ 0.1393 USD

Table 4.3 Respondent attributes and percentages (N = 580)

Attribute Item: Number of responses (%)	Attribute Item: Number of responses (%)
Gender Male 267 (46.03) Female 313 (53.97)	**Residential region:** Rural 81 (13.97) Urban 499 (86.03)
Age 18–29 years 317 (54.66) 30–39 years 177 (30.52) 40–49 years 46 (7.93) 50–59 years 26 (4.48) 60 or older 14 (2.41) **Marital status** Unmarried 281 (48.45) Married 299 (51.55)	**Occupation** Food industry worker 14 (2.41) Civil servant 50 (8.62) General corporate employee 223 (38.45) Researcher 30 (5.17) Media professional 10 (1.72) Student 156 (26.9) Farmer 13 (2.24) Migrant worker 20 (3.45) Self-employed 45 (7.76) Unemployed 19 (3.28)
Annual income 20,000 CNY or less 156 (26.9) 20,000–50,000 CNY 152 (22.21) 50,000–100,000 CNY 165 (28.45) 100,000 CNY or more 107 (18.45)	**Educational background** Junior high school and below 26 (4.48) High school or secondary school 54 (9.31) University or college of technology 264 (45.52) Graduate school 236 (40.69)

Source Created by the author based on survey results

Table 4.4 List of observed variables in the model (N = 580)

Classification	Variable		Parameter description	Mean	± SD
Endogenous observation variables	High reliability Medium reliability Low reliability	Y_1 Y_2 Y_3	1 = Strongly disagree, 2 = Disagree 3 = Neither, 4 = Agree 5 = Strongly agree	2.0 3.25 3.5	1.01 0.98 0.96
Exogenous observation variables	Reliability of internet information	X_1	1 = Very unreliable, 2 = Unreliable 3 = Neither, 4 = Reliable, 5 Very reliable	2.84	1.02
	Reliability in government HP (GHT)	X_2	Same as above	3.37	1.274
	Reliability in corporate HP (KHT)	X_3	Same as above	2.63	1.037
	Reliability in news portals (NHT)	X_4	Same as above	3.34	0.934
	Reliability in related association websites (AHT)	X_5	Same as above	3.33	1.084
	Reliability in BBS (FHT)	X_6	Same as above	2.97	1.062
	Reliability in blogs (BHT)	X_7	Same as above	2.82	1.023
	Reliability in Weibo (WBT)	X_8	Same as above	2.92	0.997
	Reliability in Wechat (WXT)	X_9	Same as above	2.83	1.057
	Reliability in information from government	X_{10}	Same as above	3.45	1.207
	Reliability in information from businesses	X_{11}	Same as above	2.54	1.008
	Reliability in information from related associations	X_{12}	Same as above	3.17	1.037
	Reliability in mass media information	X_{13}	Same as above	3.18	0.932
	Reliability in information from experts	X_{14}	Same as above	3.06	1.111
	Reliability in information from relatives and friends	X_{15}	Same as above	3.35	1.015
	Reliability in information from celebrities	X_{16}	Same as above	3.04	0.970
	Reliability in information from doctors	X_{17}	Same as above	3.71	1.005

Source Created by the author based on survey results

Table 4.5 Reliability statistics

Cronbach Alpha	Number of items
0.892	17

Table 4.6 KMO and Bartlett tests

Kaiser-Meyer-Olkin measure of sample validity		0.862
Bartlett sphericity test	Approximate Chi-square	4945.504
	degree of freedom	413
	Significance probability	0.000

test ($x^2 = 4945.504$, $p = 0.000 \leq 0.001$). Since there is a significant relationship between the indices, it can be said that it is reasonable to perform factor analysis.

b. Principal component analysis

In this study, as shown in Table 4.5, three factors were extracted from 12 items by the principal component analysis method. The eigenvalues and contributions of the first factor are 6.16 and the contribution is 38.5%, the eigenvalues of the second factor are 2.52 and the contribution is 15.75%, and the eigenvalues of the third factor are 1.19, the contribution is 7.44%, and the cumulative contribution is 61.69%. The factor loadings of the three variables of "reliability in BBS (FHT)", "reliability in Weibo (WBT)", and "reliability in WeChat (WXT)" are large, so the first factor was extracted and named as "reliability in social media (SNS)." In addition, the second factor was labeled "Trust in External Information," as the factor loadings were high for four variables: trust in information provided by mass media, experts, celebrities, and physicians. The third factor was labeled "Trust in Supervisory and Regulatory Information," as five variables—trust in government websites (GHT), corporate websites (KHT), information provided by the government, information provided by business operators, and information provided by industry associations or third-party organizations—loaded highly on this factor. In this study, mass media refers to conventional media such as newspapers and television, and is distinguished from social media (SNS) such as Weibo and WeChat (Table 4.7).

c. Construction and verification of a psychological factor evaluation model for the reliability of food safety information by covariance structure analysis

Based on the results of the above factor analysis (principal component analysis), we constructed a psychological factor evaluation model (Fig. 4.9) for the reliability of food safety information by covariance structure analysis. In addition, in the analysis, the fit was improved by removing non-significant paths and correlations, and the model with the best fit was adopted as the estimation result. The empirical model and its estimation results are outlined below. In Amos 19.0 (Japanese version), from the evaluation of the entire model, the $\chi 2$ test is $\chi 2/\mathrm{df}$ (degrees of freedom) $= 3.10$; Goodness-of-fit indices GFI $= 0.992$ (reference values are GFI > 0.9), AGFI $= 0.881$ (reference values are $0.8 \sim 0.9$, and GFI$\geqq$AGFI), RMR $= 0$. Since 051 (the reference

Table 4.7 Component matrix after rotation

Observation variables	Factors		
	1	2	3
FHT	**0.752**	0.220	0.047
WBT	**0.892**	0.129	0.124
WXT	**0.743**	0.266	0.047
GHT	−0.224	0.401	**0.695**
KHT	0.207	−0.002	**0.814**
Reliability in information from government	−0.147	0.540	**0.645**
Reliability in information from businesses	0.238	0.033	**0.752**
Reliability in information from related associations	0.209	0.444	**0.588**
Reliability in mass media information	0.450	**0.527**	0.236
Reliability in information from experts	0.120	**0.613**	0.308
Reliability in information from celebrities	0.353	**0.605**	0.186
Reliability in information from doctors	0.141	**0.766**	0.067

Note 1 The factor extraction method uses principal component analysis. The rotation method is a varimax method with Kaiser's normalization

Note 2 For observation variables, only the absolute value of factor loading after varimax rotation is 0.5 or more

value is closer to 0, the better the model fits the data), and RMSEA = 0.069 (the reference value is 0.05 ~ 0.08), the goodness-of-fit index of the model is considered to be reasonable.

Furthermore, as shown in Table 4.8, the relationship between the observed variables and the latent variables of the model is examined by partial evaluation of the model. First, a t-test is used to test the significance of the path coefficients between latent variables. The path coefficients for "reliability in SNS" and "reliability in food safety information" were not statistically significant compared to the results of the t-test. In this regard, the current spread of social media does not necessarily increase the reliability in food safety information. Information inundation can always have a negative effect on consumers' reliability in food safety information. In other words, due to the spread of the Internet and the biased public interest in information on the functionality of food ingredients, which can be called food faddism, information freely disseminated by individuals on SNS, so food information whose authenticity is unknown and information whose reliability is not guaranteed, has been flooded. This means that it may be difficult to distinguish it from reliable information. Food-faddism is defined as "the exaggerated belief in the effects of food on health and disease" or "an abnormal pattern of eating behaviour enthusiastically adopted by its adherents" (McBean, 1974). Regarding its effects, it is also noted that it can incite food insecurity.

In addition, "reliability in information from monitoring and management sectors" and "reliability in external information" were statistically significant, and "reliability

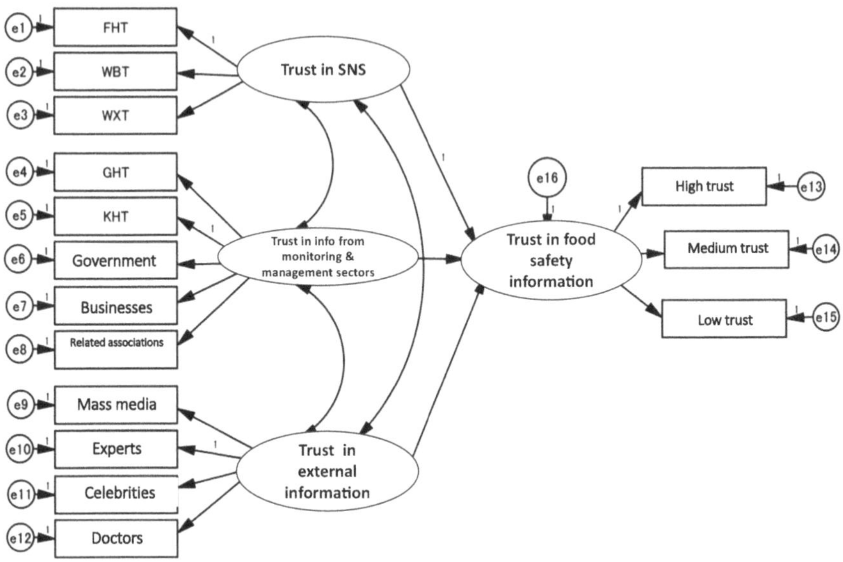

Fig. 4.9 Psychological factor evaluation model for reliability in food safety information

in information from monitoring and management sectors" and "reliability in food safety information" had a strong positive correlation (standardization coefficient of 0.775), while there was a strong negative correlation with "reliability in external information" including information from mass media. As a result of t-testing, the path coefficients of the observed and latent variables, it was found that there was a positive correlation between the nine observed variables and the three latent variables, all of which were significant at the level of $p < 0.001$.

As shown in Table 4.9, from the results of the measurement of exogenous latent variables, the path coefficients for "reliability in SNS" and "reliability in external information" are significant at the $p < 0.001$ level, and the path coefficients for "reliability in information from monitoring and management sectors" and "reliability in external information" are also significant at the t-test. There is a strong positive correlation between the variables that were significant. However, although the path coefficients for "reliability in SNS" and " reliability in information from monitoring and management sectors" were significant in the t-test, the correlation between them was very weak. In this regard, rather than denying the relationship between the use of social media and consumers' reliability in food safety information (positive correlation), it is considered that there are issues with the inadequacy and inefficiency of the use of social media in food safety monitoring and management and information disclosure in China, or in the operation and management of social media.

Table 4.8 Model measurement results

Path coefficient	Standard error	Standardized coefficients	t-test	P-value
Reliability in food safety information <—Reliability in SNS		0.942	1.434	0.18
Reliability in food safety information <—Reliability in information from monitoring and management sectors	0.142	0.775	7.611	***
Reliability in food safety information <—Reliability in external information	0.141	−0.881	−7.520	***
WBT < —Reliability in SNS	0.069	0.827	16.546	***
WXT < —Reliability in SNS	0.070	0.761	15.730	***
GHT < —Reliability in information from monitoring and management sectors	0.146	0.833	13.152	***
Government <—Reliability in information from monitoring and management sectors	0.146	0.908	13.505	***
Businesses <—Reliability in information from monitoring and management sectors	0.094	0.503	9.726	***
Related associations <—Reliability in information from monitoring and management sectors	0.101	0.582	10.755	***
Mass media <—Reliability in external information	0.079	0.686	12.574	***
Celebrities <—Reliability in external information	0.080	0.622	11.767	***
Doctors <— Reliability in external information	0.081	0.575	11.121	***

Note 1 *** $P < 0.001$; If the absolute value of the t-value is greater than or equal to 1.96, it is significant at the 5% level ($p < 0.05$) in the two-sided test; and if the absolute value of the t-value is greater than or equal to 2.58, it is significant at the 1% level ($p < 0.01$) in the two-sided test
Note 2 Only statistically significant results of t-tests on the path coefficients of the observed variables and latent variables are described

4.4.4 Correlation Analysis Between Reliability in Food Safety Information and Demographic Variability

Previous studies have pointed out that there are significant differences in the way consumers think about risk and their tolerance depending on their socioeconomic status (residential area, income, etc.), age, educational background, and so on (Slovic, 1987). In other words, differences in risk perception ability of different consumer groups may affect their responses to food safety information. Based on the results of

Table 4.9 Measurement results of exogenous latent variables

Path coefficient	Standard error	Standardized coefficients	t-test	P-value
Reliability in SNS <—> Reliability in external information	0.039	0.782	9.241	***
Reliability in SNS <—> Reliability in information from monitoring and management sectors	0.020	0.149	2.956	0.003
Reliability in information from monitoring and management sectors <—> Reliability in external information	0.030	0.696	8.107	***

Note *** $p < 0.001$

this survey, a regression analysis was conducted to clarify the relationship between the attributes of the survey respondents (see Table 4.10) and the reliability in food safety information. The results of the analysis are shown in Table 4.11.

As shown in Table 4.11, the regression analysis results show that the personal attributes of consumers have a relatively large potential to affect the "medium reliability" of food safety information. Specifically, in the case of "low reliability" in food safety information, only the consumer's gender and educational background are influencing factors, and there is a weak negative correlation. In the case of "medium reliability", it is influenced by the consumer's gender, age, family, educational background, annual income, etc. There is a weak positive correlation between the reliability of food safety information and the age and educational background of consumers, and a relatively strong positive correlation with gender and the presence or absence of elderly person(s) in the household. In addition, educational background has little effect on "medium reliability". In the case of "high reliability", the influence of educational background is stronger, and there is a positive correlation with the presence or absence of elderly person(s). There is no correlation between "low reliability" and "high reliability" for disposable income, and there is a weak negative relationship with "medium reliability".

The following points became apparent from these analytical results:

1. Female tend to transition from "low reliability" to "medium reliability" more quickly than male, and also exhibit a higher proportion demonstrating "medium reliability". This may be attributed to female generally being more emotionally driven than male, making them relatively quicker to restore trust in food safety information;
2. As the respondents' income increases, the range of food choices expands, yet concerns about food safety also tend to rise, potentially affecting trust in food safety information;
3. When consumers possess higher educational background, they not only acquire specialised knowledge but are also encouraged to conduct their own research.

Table 4.10 List of variables for investigated attributes (N = 580)

Variable	Variable descriptions	Mean	± SD
Gender	1 = Male, 2 = Female	1.54	0.499
Age	1 = 19–29 years, 2 = 30–39 years, 3 = 40–49 years 4 = 50–59 years old, 5 = 60 years old and over	1.69	0.965
Residential area	1 = Rural, 2 = Urban	1.87	−0.338
Marital status	1 = Unmarried, 2 = Married	1.52	0.5
Have child(s) or not	0 = No, 1 = Yes	0.22	0.415
Have pregnant person or not	0 = No, 1 = Yes	0.03	0.174
Have elderly person(s) or not	0 = No, 1 = Yes	0.31	0.464
Educational background	1 = Junior high school or below, 2 = High school or equivalent 3 = University or technical college, 4 = Graduate school or higher	3.22	0.794
Disposable income	1 = less than 20,000 CNY, 2 = 20,000–50,000 CNY 3 = 50,000–100,000 CNY, 4 = 100,000 CNY or above	2.38	1.07
Occupation	1 = Food business, 2 = Civil servants, 3 = Corporate employee 4 = Researcher, 5 = Media worker, 6 = Student, 7 = Farmer, 8 = Migrant worker, 9 = Self-employed, 10 = Unemployed	4.72	2.326

Consequently, their cognitive abilities regarding risk perception, decision-making, and information processing tend to be relatively higher, leading to increased trust in food safety information;

4. Regarding cohabitation with elderly, consumers living with older adults are more likely to harbour concerns and distrust regarding food safety, which is thought to influence their trust in food safety information.

However, no clear relationship was identified in this regression analysis between factors such as residential area, marital status, presence of infants, occupation, and trust in food safety information. Nevertheless, these are likely to be important consumer characteristics and should be addressed in future research.

Table 4.11 Reliability of food safety information by respondents' attributes

Independent variables	Regression analysis result	Dependent variable		
		Low reliability	Medium reliability	High reliability
Gender	Beta/r^2	**−0.003*/−0.006**	**0.104*/0.102**	0.058/0.061
Age	Beta/r^2	0.029/0.092	−0.023/	−0.013/−0.054
Residential area	Beta/r^2	−0.024/−0.066	0.011	0.010/0.076
Marital status	Beta/r^2	0.024/0.086	0.004/0.014	0.046/0.032
Have child(s) or	Beta/r^2	−0.017/0.006	0.017/0.033	−0.019/0.004
not	Beta/r^2	0/0	−0.004/	−0.004/0.013
Have pregnant	Beta/r^2	0.011/0.031	−0.008	**−0.083*/−0.066**
person or not	Beta/r^2	**−0.129*/−0. 149**	**−0.118**/**	**0.211***/0.192**
Have elderly	Beta/r^2	−0.002/−0.005	**−0.129**	−0.033/0.021
person(s) or not	Beta/r^2	−0.041/−0.029	**0.153**/**	−0.016/−0.042
Educational			**0.083**	
background			**−0.088*/**	
Disposable			**−0.029**	
income			−0.022/	
			−0.032	
Occupation				

Note *$p < 0.05$, **$p < 0.01$, ***$p < 0.001$

4.5 Conclusions and Discussion

The following conclusions are drawn from an empirical study using questionnaire survey data conducted in China. First, the report clarifies the status of consumer reliability in food safety information in China.

Mobile phones/smartphones and computers are the main medium for communicating food safety information, while SNS such as NH, WeChat and Weibo are the main sources of food safety information. Consumers also have the highest trust in food safety information from GH and relatively more trust in information from NH and AH. The amount of information from SNS has increased rapidly, but its reliability has not kept pace with the increase in the amount of information. Therefore, due to the rapid development of SNS, it is necessary for government agencies, local governments, third-party organizations, and businesses to make appropriate and effective use of SNS. Consumers have less access to information from CH, and they have the lowest trust in food safety information. In response to this, food business operators should seriously engage in risk communication to consumers regarding the disclosure of food safety information on NH and social media, the response to food safety issues, and the correction of the company's image. In addition, the older the person, the more likely they are to obtain food safety information from GH, while the less they can obtain it from social networking sites such as WeChat and Weibo. Therefore, it is considered necessary to provide effective information through a combination of websites so as to convey appropriate information according to the characteristics of each age group.

As the main source of food safety information, information from doctors, the government, and relatives and friends has gained a high level of trust, while the reliability in business is the lowest. In addition, a survey on the reliability in food safety information revealed that when food safety problems occur, there is a tendency to have "medium reliability" and "low reliability" for food safety information, and few people have "high confidence". In other words, when a person who has experienced a food safety issue or is aware of a food safety problem by many consumers, the reliability of all food safety information decreases, regardless of the source, mediator, or content of food safety information. It is believed that the situation and characteristics of consumers' low reliability of food safety information are always closely linked to consumers' overreaction to food safety issues and reputational damage. In addition, if you want to gain trust from the consumer's point of view in the current situation where "many respondents do not have 'empathy' as a recipient" regarding the low reliability of food safety information, it is extremely important to put yourself in the consumer's shoes and transmit information that can be "empathetic." In this regard, there is a theoretical basis for the generation of "emotional trust" mentioned above.

Next, the following conclusions are drawn from the results of verification of the psychological factor evaluation model of food safety information reliability by covariance structure analysis regarding the determinants of reliability in food safety information.

The reliability in food safety information is influenced by "reliability in information from monitoring and management sectors" (standardization coefficient = 0.775***, strong positive correlation) and "reliability in external information" (standardization coefficient = −0.881***, strong negative correlation). Disclosure of information on monitoring and management should be strengthened by the government, related associations, and third-party organizations, which are reliable sources of information. However, although the path coefficients for "reliability in SNS" and "reliability in information from monitoring and management sectors" were significant in the t-test, the correlation between them was very weak. In other words, the disclosure of food safety information by monitoring and controlling consumers' trust in food safety information may not be so strongly related to the use of social media. The weak correlation is thought to be due to the inadequacy and inefficiency of the use of SNS by monitoring and management organizations, as well as operational problems. For example, in China, government agencies at all levels use Weibo to set up their own "Government Affairs Weibo", but there is a lack of interaction with the public (general users) because it is only reprinted information from other organizations and entities. Several problems have been pointed out, such as a lack of presence and a lack of scientific background as a disseminator of information on SNS.

Regarding the method of disclosing food safety information, it is necessary to consider the negative impact of over-information on social media. The challenge of reliability in external information is to dispel consumers' distrust of food safety information. Mass media workers, experts, celebrities, doctors, etc., need to think about the method of transmission when conveying food safety information. We also need to consider consumers' attitudes and reliability with social media.

Previous studies have shown that women are more prone to anxiety than men (Remes et al., 2016). However, in the regression analysis of the reliability in food safety information and demographic variability in this study, fewer men and women have a "high reliability" and more women than men have a "medium reliability." In addition, women are more likely than men to switch from "low reliability" or women are more likely to leave the "low reliability" group. This suggests that women are more sensitive than men, and that it is relatively easy to restore trust in food safety information. Therefore, the development and strengthening of methods for communicating food safety information to individuals, associations and groups of women, for example, pregnant women's and mothers' associations, homemakers and housewives, and women in the workplace, is considered extremely important.

Combining the results of the empirical research in the previous chapter and this chapter, the correlation between consumers' ability to self-manage food safety (which is positively correlated with educational background, knowledge, or risk perception ability) and "reliability in food safety information" and "reliability in food safety" can be summarized as follows. In other words, self-management ability is strongly negatively correlated with the reliability of food safety information and food safety credibility if it is in the low-level interval (low education, low knowledge, and low risk cognitive ability), but it is positively correlated with reliability in food safety information and food safety reliability if it is in the high-level interval (with a certain level of educational background, knowledge, and risk recognition ability). Therefore, we believe that the reliability in food safety information should be enhanced because consumers can gain confidence in the food they choose by strengthening their self-management skills (knowledge, risk recognition ability, judgment ability, information processing ability, etc.). Strengthening self-management of food safety, including the dissemination of prerequisite scientific knowledge and careful information gathering, will curb consumers' overreaction to food risks and affect their food purchasing behavior, making it easier to purchase only reliable foods. The development of food education activities for consumers is considered to be of great significance.

Although the regression analysis does not show a relationship between the reliability of food safety information by age and region, the rapid increase in the use of the Internet and smartphones, especially highly advanced social networking sites, in rural areas and among the elderly, is likely to pose a new challenge to the reliability in food safety information.

4.6 Policy Implications: Food Safety Information Governance and Trust Formation

The findings of this chapter provide the following policy implications for the redesign of food safety governance in China.

First, it is essential to strengthen information disclosure by the government and regulatory authorities. The empirical analysis demonstrates that trust in information provided by government bodies and related institutions significantly shapes the overall reliability of food safety information. Therefore, it is necessary to ensure transparent, science-based information disclosure and to fulfill ongoing accountability.

Second, a qualitative transformation of risk communication is required. Rather than merely providing information, it is crucial to develop "empathetic communication" that takes the consumer's perspective into account. Particularly in food safety issues that easily generate anxiety and misunderstanding, it is important to utilize highly trusted actors—such as experts, medical professionals, and third-party organizations—and to promote interactive, two-way communication through diverse channels.

Third, institutional and operational improvements in the use of social media are needed. This study shows that the spread of social media (SNS) does not necessarily contribute to enhancing trust. This is attributable to factors such as information overload, the spread of unverified information, and non-interactive communication practices by public institutions. Going forward, strategic use of SNS and the establishment of appropriate guidelines for trust-building are required.

Fourth, it is necessary to strengthen corporate trust-recovery strategies. Trust in business operators remains low, and once a food safety incident occurs, distrust tends to spread across the entire industry. Companies should not rely solely on reactive crisis management but should proactively build trust through regular information disclosure, enhanced transparency, and the implementation of corporate social responsibility (CSR).

Fifth, improving consumers' risk perception capacity and information literacy is of critical importance. The analysis indicates that higher levels of education and knowledge are associated with greater trust in food safety information. Therefore, enhancing consumers' self-management capabilities—such as risk assessment, information processing, and decision-making—through food education and scientific literacy initiatives can help reduce overreaction and mitigate reputational damage.

In sum, food safety issues should not be understood merely as regulatory or technical problems but as social challenges shaped by the interaction of information, trust, and behavior. Accordingly, a comprehensive policy approach that integrates institutional design, the information environment, and consumer capability is required.

Acknowledgements This chapter is an extensively revised, expanded, and updated English version of the author's article, "Risk Communication for Enhancing Public Trust in Food Safety in China: An Empirical Analysis Based on a Questionnaire Survey on the Communication of Food Safety Information to Consumers," published in *Social Systems Studies*, No. 33, Ritsumeikan University, 2016.

References

Akerlof, G. A. (1970). The Market for "Lemons": Quality Uncertainty and the Market Mechanism. *The Quarterly Journal of Economics*, *84*(3), 488–500. https://doi.org/10.2307/1879431

China Internet Network Information Center. (2016). The 37th statistical report on internet development in China. China Internet Network Information Center.

China Internet Network Information Center. (2025). The 56th statistical report on internet development in China. China Internet Network Information Center.

Food and Agriculture Organization of the United Nations, & World Health Organization. (2013). Frequently asked questions on avian influenza A(H7N9) and food safety. FAO & WHO.

Liu, J., & Xu, W. (2024). An exploratory qualitative analysis of public awareness, communication dynamics, and behavioral patterns regarding food safety in Mainland China. *BMC Public Health*, *24*(1), 2884. https://link.springer.com/article/10.1186/s12889-024-20382-3

McBean, L. D. (1974). Food faddism: A challenge to nutritionists and dietitians. The American Journal of Clinical Nutrition, 27(10), 1071–1078. https://doi.org/10.1093/ajcn/27.10.1071

Miyoshi, E. (2012). Safety, perception of risk and social trust in social system: The China Path to Global Food Safety. *Discussion Papers in Contemporary China Studies, Osaka University Forum on China*, *2012*(01), 1–21. https://ir.library.osaka-u.ac.jp/repo/ouka/all/22996/dp2012-1miyoshi.pdf

Remes, O., Brayne, C., Van Der Linde, R., & Lafortune, L. (2016). A systematic review of reviews on the prevalence of anxiety disorders in adult populations. *Brain and behavior*, *6*(7), e00497. https://doi.org/10.1002/brb3.497

Slovic P. (1987). Perception of risk. *Science (New York, N.Y.)*, *236*(4799), 280–285. https://doi.org/10.1126/science.3563507

Wang, F., Zhou, W., Qian, X., & Nakagami, K. (2016). Risk communication aimed at improving trust in food safety in China: An empirical analysis based on a consumer survey on food safety information dissemination. *Social Systems Studies*, *33*, 55–85. https://doi.org/10.34382/00003964

Wang, X., Xu, Y., Liang, M., Gao, J., Wang, J., Chen, S., & Cheng, J. (2022). Chinese consumers' trust in food safety surveillance sampling of commonly consumed foods. *Foods*, *11*(13), 1971. https://doi.org/10.3390/foods11131971

World Health Organization. (n.d.). Questions and answers on avian influenza A(H7N9) or "bird flu". World Health Organization.

Yamagishi, T., & Komiyama, H. (1995). Significance and the structure of trust: Theoretical and empirical research on trust and commitment relations. INSS Journal, 2, 1–59. http://www.inss.co.jp/wp-content/uploads/2017/03/1995_2J001_059.pdf

Part II
Institutions, Governance, and the Institutionalization of Trust: Institutional Trust Formation and Collaborative Governance

Chapter 5
Food Safety Systems and Trust Foundations from a Japan–China Comparative Perspective: A Systems Analysis of Improving Food Safety Reliability and Legislation in China

Fengyang Wang, Xuepeng Qian, and Weisheng Zhou

5.1 Introduction

During the 1980s, Japan had already undergone a period of high economic growth, resulting in a high nominal GDP per capita. This was underpinned by advanced industrialization and the development of a service-based economy. Food supply remained sufficient, providing approximately 2,600 to 2,800 kilocalories per person per day. Meanwhile, following its reform and opening-up policy, China achieved rapid economic development from its low-income levels in the 1980s. By the early twenty-first century, its nominal GDP per capita had exceeded the $1,000 range, reaching approximately $13,000 by 2024. This is assessed as being close to Japan's income level in the early 1990s. As shown in Table 5.1, both countries have seen an upward trend in per capita calorie supply alongside economic growth. Particularly in China, heightened concern for nutrition and food safety has progressed alongside rising living standards and dietary diversification. In Japan, high incomes and stable food supplies were achieved relatively early, allowing per capita calorie supply to be maintained at a high level over the long term. In China, however, calorie supply

F. Wang
Xi'an Peihua University, Xi'an, China

X. Qian
Graduate School of Global Environmental Studies, Sophia University, Tokyo, Japan
e-mail: qianxp@sophia.ac.jp

W. Zhou (✉)
College of Policy Science, Ritsumeikan University, Kyoto, Japan
e-mail: zhou@sps.ritsumei.ac.jp

W. Zhou (ed.), *Food Safety in China*, https://doi.org/10.1007/978-981-92-3741-8_5

has improved alongside economic development, and the nutritional status of the population and awareness of food safety are steadily increasing.

In recent years, China has achieved rapid economic development, raising living standards whilst expanding the prevalence of eating out and increasing demand for foodstuffs such as meat products, dairy products, aquatic products, coarse grain products, and canned goods. Consequently, whilst the food processing industry has developed and food distribution networks have been established, consumer concern regarding food safety has grown. The Chinese government is seriously addressing food safety issues, leading to a rapid strengthening of preventive management and responses to hazards. However, as consumer anxiety regarding food safety persists, challenges remain in China concerning the reinforcement of legal systems for food safety and security, the effective enforcement of laws, the review of responsible organizations and staffing arrangements, and particularly the promotion of risk communication as a policy tool aimed at alleviating consumer concerns and distrust (Bian, 2012; Wu et al., 2018).

On the other hand, Japan is internationally recognized as one of the countries with well-established food safety management systems and quality control frameworks. Food produced in Japan has long enjoyed relatively high regard and trust in terms of quality and safety. According to the 2014 Survey on Consumption Trends of Visitors to Japan, approximately 80% of foreign visitors cited "enjoying Japanese cuisine" as their primary purpose for visiting (Japan Tourism Agency, 2025). Although survey items and question formats differ, the Japan Tourism Agency's 2024 Inbound Consumer Trends Survey also indicates that Japanese cuisine remains one of the primary motivations for visiting Japan, with approximately 65% of foreign visitors stating they wish to enjoy Japanese food on their next trip (Japan Tourism Agency, 2025).

According to the Consumer Trends Survey for the First Half of Fiscal Year 2014 conducted by the Japan Finance Corporation's Agriculture, Forestry and Fisheries Division in July 2014, a survey on domestic and imported food products revealed that 77.4% of respondents stated they "pay attention" to whether food products are domestically produced when purchasing groceries. Furthermore, 69.7% reported having the impression that food made from domestic ingredients is "safe" (Japan Finance Corporation, 2014).

Table 5.1 Comparison of nominal GDP per capita and calories per capita (food supply) between Japan and China

Indicator	China		Japan	
	1980	2024	1980	2024
GDP per capita	300 USD (low income)	13,000 USD	8,900 USD	33,800 USD
Calories per capita per day	2,100–2,300 kcal	2,800–3,000 + kcal	2,800–3,200 + kcal	3,000–3,500 + kcal

Source Food and Agriculture Organization of the United Nations (2024),World Bank (2026)

Meanwhile, the latest survey, the Japan Finance Corporation's Consumer Trends Survey (July 2024), indicates that the proportion of consumers who stated they "pay attention" when purchasing domestic food products (including those containing domestic ingredients) remained at a high level, ranging from approximately 66.3% to 73.2%. It also shows that a significant majority of consumers, approximately 63.4%, hold the perception that domestic food products are "safe". This proportion is reported to show a slight upward trend compared to the previous survey. Regarding imported food, however, the proportion answering that there are safety concerns' was approximately 31.8%, indicating a tendency to view them with greater apprehension compared to domestic products (Japan Finance Corporation, 2024).

These survey results indicate that consumers continue to hold a strong perception of safety and security regarding domestically produced food.

The primary reason for this lies in the fact that, during its post-war economic development, Japan utilized the lessons learned from numerous food safety incidents and accidents to enact and amend legislation, while also establishing various standards and criteria for each stage of food production, processing, distribution, and consumption. Furthermore, the establishment of the Consumer Affairs Agency, the introduction of risk analysis, and the promotion of food education have built a food safety administration system demanding a "consumer-first" approach and "preventive action". For China, where consumer anxiety and distrust regarding food safety are becoming apparent, drawing on Japan's experience in its legal framework for food safety, the management system of administrative bodies, and particularly the promotion of consumer-oriented food safety risk communication, is considered beneficial.

The validity of the comparison is fully mindful of the differing political systems and economic circumstances of the two countries. However, given the considerable similarity in agricultural types, staple foods, and dietary habits, both governments share a background of long-term emphasis on food safety issues. Furthermore, as mentioned earlier, commonalities can be discerned even in food safety issues arising under differing stages of development. Based on these similarities, comparative research will proceed.

The theoretical and empirical analyses in this book elucidate the theoretical and practical importance of deepening clear and efficient risk communication among all stakeholders and consumers within contemporary food safety systems to foster and enhance trust. Furthermore, it identifies the current state and characteristics of trust in food safety and trust in food safety information. Through analysis considering consumer attributes, it presents policy directions for constructing food safety systems that contribute to enhancing food safety trustworthiness and for effective risk communication approaches.

This chapter focuses on food safety management systems from a comparative Japan–China perspective. Following an institutional comparison of food safety assurance systems in both countries, it aims to further examine the current state and challenges of consumer-oriented risk communication in China, along with future initiatives.

5.2 Institutional Comparison of Ensuring Food Safety in Japan and China

5.2.1 Laws and Administrative Systems for Ensuring Food Safety in Japan

In the context of significant food safety concerns, legislative measures have been implemented and amended to guarantee food safety in Japan, accompanied by the refinement of the administrative system. In order to understand the process of formulating and developing these legal and administrative systems, it is necessary to first describe the changes in the people's dietary habits and the shift in their awareness of food safety.

5.2.1.1 Changes in Dietary Habits and Shifts in Food Safety Awareness: Towards an Emphasis on Qualitative Safety

Since the 1970s, Japan's GDP per capita (purchasing power parity, PPP) has increased substantially alongside economic growth. By 2024, it had reached approximately US$46,097, representing a 14-fold increase from around US$3,300 in 1970. Throughout this period, Japan has maintained a high economic standard globally, and the affluence of its citizens has improved dramatically.

On the other hand, per capita energy intake has not necessarily increased in proportion to economic growth; rather, a declining trend is observed. According to the National Health and Nutrition Survey conducted by the Ministry of Health, Labour and Welfare, the average energy intake in the 2020s was approximately 1,900 kcal/day, representing a decrease compared to the peak levels of the 1970s (Fig. 5.1). Factors contributing to this background are thought to include changes in lifestyle, the Westernisation of food composition, and population ageing.

In other words, it can be said that dietary habits have changed alongside economic development. Moreover, as shown in Table 5.1, alongside economic development and dietary changes, public attitudes towards food have also undergone significant transformation. Whereas securing sufficient food quantities was once the primary concern, present-day focus has shifted towards nutritional balance and food safety. In essence, the concept of food security has broadened and deepened from "quantity" to "quality". This reflects how, alongside the affluence brought by economic development, health consciousness and awareness of food safety have permeated society as a whole.

The broadest concept of "food security" for Japan addresses the challenge of achieving stable international food supplies and raising Japan's food self-sufficiency rate. Next, "food safety" is understood as ensuring the safety of food concerning issues such as food poisoning, pesticide residues, food additives, and genetically modified crops, matters currently addressed by the Food Safety Commission.

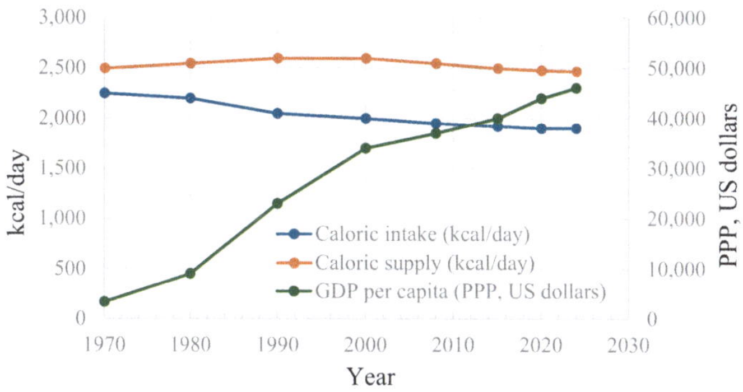

Fig. 5.1 Economic level and changes in food consumption. *Source* Per capita GDP (PPP) is based on OECD (2024), and per capita caloric intake and supply are created by the author based on FAO (2024) data

In particular, as shown in Table 5.2 and Fig. 5.1, the spread of the COVID-19 significantly impacted the dietary habits of the Japanese population. Restrictions on going out and the widespread adoption of working from home increased opportunities for cooking at home, leading to a higher proportion of meals eaten at home. Conversely, the use of eating out and ready-made meals temporarily decreased, while takeaway and delivery services saw increased utilization, particularly due to restrictions on restaurant operations. Besides, food purchasing behavior also changed, with an increase in bulk buying and purchases of foods with a longer shelf life. This reflects consumer behavior influenced by concerns over supply chain instability and infection risk. Furthermore, awareness regarding food safety shifted, with heightened interest not only in food hygiene but also in domestic production and the stability of supply. The importance of hygiene management and reliable food supply chains was reaffirmed.

Furthermore, COVID-19 has catalyzed the establishment of a hybrid approach combining home cooking, eating out, and ready meals, fostering flexible and diverse dietary patterns. Health consciousness and criteria for food selection are increasingly being reassessed, with greater emphasis placed on balancing hygiene, safety, and convenience. Thus, COVID-19 has brought about significant changes not only in the form of dietary habits but also in attitudes towards food safety.

The overall food self-sufficiency rate is an indicator showing the self-sufficiency rate for all food. It is calculated using two methods: on a supply calorie basis and on a production value basis. The calorie-based comprehensive food self-sufficiency rate is calculated by converting the weight of each item into supply calories based on the "Japanese Food Composition Tables 2010", then summing the values. The production value-based comprehensive food self-sufficiency rate is calculated by converting the weight of each item into monetary value based on farm-gate prices from the "Agricultural Price Statistics", then summing the values. Furthermore, Japan's overall food self-sufficiency rate has been significantly influenced by major changes in food supply and demand, such as the Westernization of dietary habits during the period

Table 5.2 Changes in dietary habits and food safety by time in Japan

Time period	Changes in dietary habits and food safety
Postwar period (1945–1955)	(1) Food shortages; (2) In order to increase production, global environmental problems and issues of ensuring the "safety and security" of food have emerged due to the increase in agriculture with heavy use of pesticides and chemical fertilizers and livestock farming; (3) Yellow rice found in imported rice; (4) Emphasis on standardization and quality labeling of feed, livestock hygiene, agricultural and forestry materials; (5) Establishment of the Food Sanitation Act and development of food sanitation administration
Period of high economic growth (1956–1973)	(1) Worsening pollution issues and enactment of related laws; (2) Westernization of food culture, rapid decline in food self-sufficiency, and manifestation of overproduction of rice; (3) The emergence of frozen foods and food service, and the expansion of food industry; (4) Contamination of hazardous chemical substances in food factories as a social issue
Period of low economic growth (1974–1984)	(1) Simplification of food, gourmet food, and first advocacy of Japanese-style eating habits; (2) Huge growth in the foodservice and catering industry; (3) Increase in obesity and diabetes
Bubble Economy Period (1985–1990)	(1) Trade friction and liberalization pressure; (2) Globalization of food industry and safety issues of imported foods emerge; (3) Rapid increase in food distribution at the individual level
Period of long-term stagnation (After the Bubble Economy Period) (1991–2010)	(1) In 2001, BSE-infected cattle were discovered, which had a significant impact on meat consumption; (2) Mislabeling of the origin of beef by a major food manufacturer was discovered. Subsequently, cases of mislabeling of food products surfaced one after another; (3) Implementation of the Food Safety Basic Act and transition to Administration of Food Safety
After the Great East Japan Earthquake in 2011	(1) Emergency food security and food safety revisited; (2) Spread of radioactive materials and food safety issues
Pre-COVID-19 (by 2019)	The expansion of eating out and ready meals accelerated the externalization of food, with convenience and diversity becoming paramount. Meanwhile, concerns regarding food safety were primarily directed towards qualitative safety aspects such as food hygiene, labelling, and origin

(continued)

Table 5.2 (continued)

Time period	Changes in dietary habits and food safety
During COVID-19 (2020–2022)	The increase in home cooking resulted from reduced outings and the expansion of remote working, temporarily boosting cooking opportunities at home. Food purchasing shifted towards bulk buying and prioritizing shelf life. Concerns over distribution disruptions and reliance on imports heightened awareness of the importance of stable food supply and domestic production. Food safety came to be viewed not only in terms of hygiene management but also from the perspective of supply reliability
Post-COVID-19 (from 2023)	Dining-out demand is gradually recovering, yet the practice of combining it with home-cooked meals and takeaway food has become established, leading to greater diversity in dietary habits. Regarding food safety, against the backdrop of heightened hygiene awareness stemming from infection control measures, a comprehensive perspective encompassing food safety, stable supply, and sustainability is increasingly being prioritized

Source Compiled by the author based on Ministry of Agriculture, Forestry and Fisheries (2021) and related food safety laws and regulations

of high economic growth. The calorie-based food self-sufficiency rate declined from approximately 73% in 1965 to around 37% in 2020, while the production value-based rate also decreased from approximately 86% to around 65% over the same period. Thus, alongside economic development, the food self-sufficiency rate has significantly fallen, making the securing of stable food supply an increasingly critical challenge (Ministry of Agriculture, Forestry and Fisheries, 2021).

It remains necessary to assume a certain volume of food imports going forward, and given heightened consumer concern regarding the safety of imported foodstuffs, ensuring and promoting their safety is a critical challenge. Consumer awareness of food safety has been further heightened by incidents such as the 1996s 0157 food poisoning outbreak (over 6,000 patients), the 2000 Snow Brand processed milk food poisoning incident (over 14,000 patients), and the 2001 BSE incident. As illustrated in Fig. 5.3, a 2002 survey on consumer trends suggests that one background factor for enacting the Basic Act on Food Safety and revising the Food Sanitation Act was that consumer awareness and concern regarding food safety had surpassed mere quantitative assurance. Furthermore, the frequent occurrence of food safety issues has heightened public distrust in food safety. Having grasped these subjective and objective factors, the process of forming and perfecting Japan's legal and administrative framework for food safety are summarized.

Moreover, as shown in Fig. 5.2, the externalization of food consumption is progressing against a backdrop of women entering the workforce, an increase in single-person households, a declining birth rate and ageing population, and diversifying lifestyles. The consequences include nutritional imbalances, irregular eating patterns, rising obesity rates, increased lifestyle-related diseases, excessive slimming

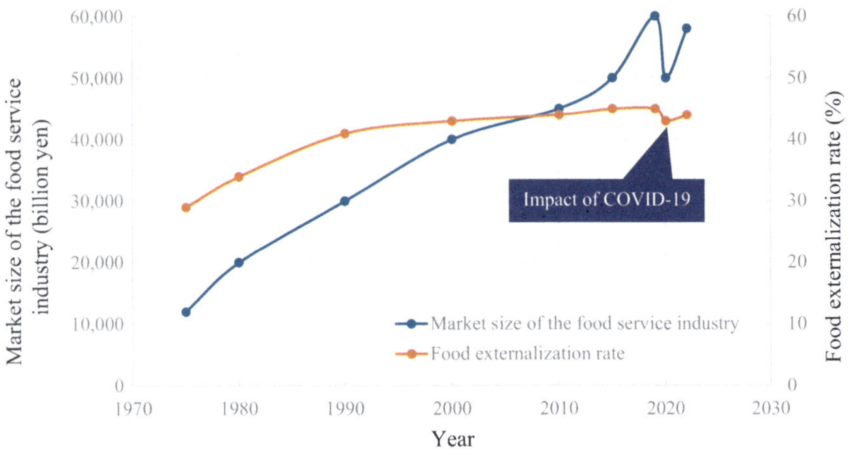

Fig. 5.2 Changes in the rate of home-meal replacements and the market size of foodservice industry in Japan. *Note 1* Against a background of increasing dual-career and single-person households, an ageing population and diversifying lifestyles, there is a growing dependence on home-meal replacements and eating outside the home. In line with this, progress is being made in the food industry in providing and developing markets for "ready-made meals", such as ready-made food, side dishes and packed lunches, in response to changes in food consumption patterns. These trends are collectively referred to as "home-meal replacement" (Ministry of Agriculture, Forestry and Fisheries, 2013). *Note 2* The rate of home-meal replacements is the proportion of ready-made food and eating out as a percentage of total food expenditure. *Source* Created by the author based on the Japan Food Service Association (2023) and the Ministry of Agriculture, Forestry and Fisheries (2021)

trends, and frequent large-scale food poisoning incidents stemming from the rapid development of the food service industry and mass production with extensive distribution networks. Amidst this, concern for food safety has also grown. Preventing the occurrence of food poisoning and mitigating its spread are crucial challenges for legislation and administrative systems.

5.2.1.2 Overview of the Legal System for Food Safety and Its Characteristics

With the post-war economic development, global environmental problems due to the increase in agriculture and livestock farming, which use a lot of pesticides and chemical fertilizers, and the issue of ensuring the "safety and security" of food have emerged. Therefore, a series of laws such as the Agricultural Chemicals Regulation Act (1948), the Act on Japanese Agricultural Standards (1950), and the Slaughterhouse Act (1953) were enforced in Japan. Subsequently, incidents such as falsification of production places and expiration dates, and unauthorized addition of chemical substances occurred, and some laws were amended and new laws were enacted, and a legal system for ensuring food safety at all stages of food production, food processing,

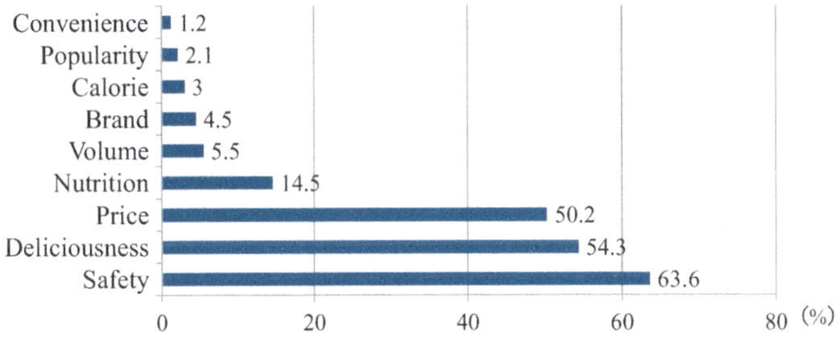

Fig. 5.3 Consumer awareness and concerns when purchasing food in Japan. *Source* Created by the author based on data from the Agricultural, Forestry and Fisheries Finance Corporation, "2002 First Survey on Consumer Trends, etc. (Questionnaire Survey on Food Labeling)" (survey conducted June–July 2002)

and food distribution was established. The current system of laws for ensuring food safety in Japan contains more than 300 laws, regulations, and notices, as organized in the Food Sanitation Statute Handbook (2014 Edition). Among them, the composition of the main laws is summarized and shown in Table 5.3.

As shown in Table 5.3, Japan's general legal framework aimed at ensuring food safety shows the following three characteristics.

First, the legal framework has a broad scope, which is in place at every stage of the "from farm to table" process, including agricultural production, food production and processing, and food distribution and importation.

Second, the laws are revised and redeveloped in a timely and appropriate manner. In other words, relevant legislation is revised and supplemented promptly in response to changes in food safety-related situations and circumstances. For example, the Food Sanitation Act forms the basis of food sanitation administration and has been amended 32 times since its enactment on 24 December 1947. The main provisions have changed from Article 33 to Article 79, and the supplementary provisions have also been significantly revised. The main provisions of the law stipulate "substantive provisions that form the main part of the law," and the supplementary provisions stipulate "matters that are necessary incidental to the matters stipulated in these rules." In addition, the Food Safety Basic Act, enacted in 2003, was amended in 2007 (this amendment did not involve major changes to fundamental principles or the framework, but rather adjusted the wording of articles and ensured consistency with related legislation). Under such circumstances, provisions for specifications, classification, standards and labeling of food additives have been developed in a timely and appropriate manner. An example is the classification of types and uses of food additives under the Food Sanitation Act. Additives other than those designated under Article 10 of the Food Sanitation Act (designated additives) may not be used, but natural additives listed on the list of existing additives are allowed to continue to be used even if they are not designated (Article 2 and 3 of the Supplementary Provisions of the Food Sanitation Act [Act No. 101 of May 24, 1995]). The revision in 2003 added a provision that allows existing additives to be removed from the

Table 5.3 Major laws and regulations related to food safety in Japan (with partial overlap depending on classification)

	The name of the law (the time of enactment and revision)
General laws related to food safety	(1) Food Safety Basic Act (Act No. 48 of 2003, revision in 2009) (2) Food Sanitation Act (Act No. 233 of 1947, major revision in 2003, revision in 2013)
Laws related to agricultural production	(1) Livestock Hygiene Service Centers Act (Act No. 12 of 1950, revision in 1999) (2) Act on the Prevention of Infectious Diseases in Livestock (Act No. 166 of 1951, revision in 2013) (3) Act on Stabilization of Production and Shipment of Vegetables (Act No. 103 of 1966, revision in 2013) (4) Act to Prevent Soil Contamination on Agricultural Land (Act No. 139 of 1970, revision in 2011) (5) Act on Stabilization of Supply, Demand and Prices of Staple Food (Act No. 113 of 1994, revision in 2011) (6) Basic Act on Food, Agriculture and Rural Areas (Act No. 106 of 1999, revision in 2009) (7) Act on Promotion of Introduction of Sustainable Agricultural Production Practices (Act No. 110 of 1999, revision in 2013) (8) Sustainable Aquaculture Production Assurance Act (Act No. 51 of 1999, revision in 2007) (9) Act on the Establishment of Laws Related to the Ministry of Agriculture, Forestry and Fisheries to Ensure Food Safety (enactment in 2003, revision in 2013) (10) Act on the Conservation and Sustainable Use of Biological Diversity through Regulations on the Use of Living Modified Organisms (Act No. 97 of 2003, revision in 2013) (11) Act on Promotion of Organic Agriculture (Act No. 112 of 2006, revision in 2011)

(continued)

Table 5.3 (continued)

	The name of the law (the time of enactment and revision)
Laws related to food production	(1) Agricultural Chemicals Regulation Act (Act No. 82 of 1948, revision in 1971, 2002, 2003, 2007) (2) Plant Protection Act (Act No. 151 of 1950) (3) Act on the Quality Control of Fertilizer (Act No. 127 of 1950, revision in 2011) (4) Act on Safety Assurance and Quality Improvement of Feeds (Act No. 35 of 1953, revision in 2007) (5) Slaughterhouse Law (Act No. 114 of 1953, revision in 2007) (6) Sewerage Act (Act No. 79 of 1958, revision in 2011) (7) Act on Securing Quality, Efficacy and Safety of Products Including Pharmaceuticals and Medical Devices (Act No. 145 of 1960, revision in 2013) (8) Confectionery Hygienist Act (Act No. 115 of 1966, revision in 2011) (9) Poultry Slaughtering Business Control and Poultry Meat Inspection Act (Act No. 70 of 1990, revision in 2013) (10) Act on Temporary Measures concerning Sophistication of Management of Food Manufacturing Process (or HACCP Act) (Act No. 59 of 1998, revision in 2013) (11) Act on Promotion of Recycling and Related Activities for Treatment of Cyclical Food Resources (Act No. 116 of 2000, revision in 2013) (12) Act on Special Measures concerning Measures against Bovine Spongiform Encephalopathy (Act No. 70 of 2002, revision in 2003)

(continued)

Table 5.3 (continued)

	The name of the law (the time of enactment and revision)
Laws related to the distribution and import of food products	(1) Import Trade Control Order (Cabinet Order No. 414 of 1949, revision in 2003) (2) Agricultural Products Inspection Act (Act No. 144 of 1951, revision in 2011) (3) Act on Special Measures concerning Prevention of Toxic Contamination of Food Distributed Through Marketing Channels (Act No. 103 of 1987) (4) Act on Promotion of Recycling and Related Activities for Treatment of Cyclical Food Resources (Act No. 116 of 2000, revision in 2013) (5) Beef Traceability Act (Act No. 72 of 2003, and revision in 2008) (6) Act on the Recording of Information Related to Transactions of Rice Grain and the Transmission of Information on Production Areas (Act No. 26 of 2009, revision in 2013)
Laws related to food labeling	(1) Food Sanitation Act (Act No. 233 of 1947, major revision in 2003, revision in 2013) (2) Act on Standardization and Proper Quality Labeling of Agricultural and Forestry Products (or JAS Act) (Act No. 175 of 1950, revision in 2013) (3) Measurement Act (Act No. 51 of 1992) (4) Act against Unjustifiable Premiums and Misleading Representations (Act No. 134 of 1962, revision in 2013) (5) Household Goods Quality Labeling Act (Act No. 104 of 1962, revision in 2011) (6) Act on the Promotion of Sorted Collection and Recycling of Containers and Packaging (Act No. 112 of 1995, revision in 2011) (7) Health Promotion Act (Act No. 103 of 2002, revision in 2013) (8) Beef Traceability Act (Act No. 72 of 2003, and revision in 2008) (9) Act on the Recording of Information Related to Transactions of Rice Grain and the Transmission of Information on Production Areas (Act No. 26 of 2009, revision in 2013) (10) Food Labeling Act (Act No. 70 of 2013)

(continued)

Table 5.3 (continued)

	The name of the law (the time of enactment and revision)
Laws related to the control of agricultural residues in food	(1) Agricultural Chemicals Regulation Act (Act No. 82 of 1948, revision in 1971, 2002, 2003, 2007) (2) Act on Securing Quality, Efficacy and Safety of Products Including Pharmaceuticals and Medical Devices (Act No. 145 of 1960, revision in 2013)
Regulations related to radioactive material inspections of agricultural, livestock and fishery products	(1) Concept of setting and canceling inspection plans, shipping restrictions, etc. (laws and regulations) (enactment in 2011, revision in 2012) (2) Inspection of radioactive substances in agricultural, livestock and fishery products (laws and regulations) (enactment in 2013) (3) Shipment restrictions based on the Act on Special Measures for Responding to Nuclear Disasters (enactment in 1999) (4) Restrictions on intake based on the Act on Special Measures for Responding to Nuclear Disasters (enactment in 1999) (5) Response to radioactive substances in food
Laws related to imports confirmed by customs (animal, plant, and food-related matters)	(1) Food Sanitation Act (Act No. 233 of 1947, major revision in 2003, revision in 2013) (2) Agricultural Chemicals Regulation Act (Act No. 82 of 1948, revision in 1971, 2002, 2003, 2007) (3) Import Trade Control Order (Cabinet Order No. 414 of 1949, revision in 2003) (4) Plant Protection Act (Act No. 151 of 1950) (5) Act on the Prevention of Infectious Diseases in Livestock (Act No. 166 of 1951, revision in 2013) (6) Act on the Quality Control of Fertilizer (Act No. 127 of 1950, revision in 2011) (7) Act on the Protection of Marine Resources (Act No. 313 of 1951, revision in 2010) (8) Act on Securing Quality, Efficacy and Safety of Products Including Pharmaceuticals and Medical Devices (Act No. 145 of 1960, revision in 2013) (9) Act on the Prevention of Adverse Ecological Impacts Caused by Designated Invasive Alien Species (Act No. 78 of 2004, revision in 2013)

(continued)

Table 5.3 (continued)

	The name of the law (the time of enactment and revision)
Others	(1) School Lunch Program Act (Act No. 160 of 1954, revision in 2008) (2) Act on the Regulation of Manufacture and Evaluation of Chemical Substances (Act No. 117 of 1973, revision in 2013) (3) Product Liability Act (Act No. 85 of 1994) (4) Enforcement Regulations of the Act on Special Measures against Dioxins (enactment in 1999, revision in 2010) (5) Basic Act on Consumer Policies (Act No. 78 of 1968, revision in 2012) (6) Basic Act on Shokuiku (Food and Nutrition Education) (Act No. 63 of 2005, revision in 2009)

Source Created by the author with reference to the Legal Data Provision System (Ministry of Internal Affairs and Communications, Administrative Management Bureau) and the Japan Legal Index (National Diet Library), among others

list of existing additives if there are "safety concerns" or "no actual use." In 1995, 489 items were listed in the list of existing additives, but by May 2011, 124 items had been removed. Of these removed items, the carcinogenicity of the red dye was confirmed, and the other 123 items were also removed respectively due to the lack of distribution status (Food Safety Commission of Japan, 2003; Ministry of Health, Labour and Welfare, 2023).

Finally, the third is the close interlinkage and supplementation between laws and laws, and between laws and related regulations. For example, the Act on Securing Quality, Efficacy and Safety of Products Including Pharmaceuticals and Medical Devices treats both people and veterinary products in equal measure as regulated entities. It is also noteworthy that content may be present which pertains to other legislative acts and regulations. For instance, the Act against Unjustifiable Premiums and Misleading Representations, the Measurement Act and the Health Promotion Act regulate the proper labeling of food products (Li & Nanseki, 2013). In addition, at the same time as the enforcement of the Food Safety Basic Act and the Food Sanitation Act, there are specific related regulations, standards, and norms in various fields. For example, the positive list system for agricultural residues, GAP as a risk management method for farms, HACCP certification related to food trade, and the food traceability system have been used in a coordinated manner.

5.2.1.3 Introduction of Risk Analysis and Framework of Food Safety Administrative System

(a) Background

With the outbreak of BSE-infected cattle, a "risk analysis" method related to food safety was discussed in Europe, and an organization and administrative method based on this premise were introduced. In other words, in Europe and the United States, the outbreak of BSE-infected cattle and the response to it have led to discussions centered on the concept of "risk," which indicates how likely and to what extent the health of people who consume the food will have an adverse effect on their health if there is a hazard in food. In particular, the EU's food safety authorities and other new food administrative organizations in Europe have succeeded in restoring consumer trust by drastically reviewing their existing organizations based on the basic principle of "risk analysis" in order to respond to the BSE problem.

On the other hand, in the case of Japan, a series of problems from 2001 to 2002, such as the BSE problem, false labeling, and the use of residual and unregistered pesticides in imported foods, increased public anxiety and distrust of food safety. In particular, the dairy scandal in 2000 and the BSE scandal in 2001 led to strong calls for administrative responsibility, and the government worked to reform the system to ensure food safety (Li & Nanseki, 2013). On July 1, 2003, the Food Safety Basic Act came into effect, and the basic principles are (1) protection of the lives and health of the people, (2) ensuring safety at each stage of the process of supplying food, and (3) appropriate responses in accordance with the latest scientific knowledge and international trends. However, the purpose is to "comprehensively promote policies to ensure food safety by establishing basic principles, by clarifying the responsibilities of the national and local governments, and food-related business operators and the roles of consumers, and establishing a basic direction for policy formulation, in order to ensure food safety, in consideration of the vital importance of precise responses to the development of science and technology, and to the progress of internationalization and other changes in the environment surrounding Japan's dietary habits." In addition to enhancing the solidarity between administrative agencies, securing testing and research and human resources, collecting domestic and foreign information, appropriately operating the labeling system, promoting food education, and considering the impact on the environment, the Act introduced the concept of "risk analysis" that clearly positions "food safety and security" with reference to the experience of Europe and the United States.

(b) Introduction of "risk analysis"

"Risk analysis" is a concept originally developed in the field of engineering research of economic activities in the United States, but when it is applied in the field of food safety, as shown in Fig. 5.4, is classified into three areas (1) "Risk assessment" that scientifically evaluates the effects of food consumption on human health, (2) "Risk management" in which administrative measures such as residue standards for additives and pesticides and the implementation of various regulations to ensure safety are determined based on the results of risk assessment, taking into account dietary conditions, cost-effectiveness, public sentiment, etc., and (3) "Risk communication," which involves the provision of information and the mutual exchange of opinions on both risk assessment and risk management. In July 2003, the Food Safety Commission, established in the Cabinet Office, conducted food health impact assessments and risk assessments based on scientific knowledge. Next, "risk management" implements policy measures to reduce the degree of risk at all stages of food production, distribution, and consumption. The Ministry of Health, Labour and Welfare, the Ministry of Agriculture, Forestry and Fisheries, and the Consumer Affairs Agency are the relevant administrative agencies that are responsible for the policy. Finally, risk communication involves the mutual exchange of information and opinions between the parties concerned (evaluators, managers, consumers, operators, etc.) across all stages, including assessment and management. In other words, the introduction of "risk analysis" is a process to prevent accidents and minimize risks to the extent possible, rather than to clean up after them, when a certain group of the population may be exposed to harm for the purpose of protecting the public's health.

(c) Framework of the food safety administrative system

The current food safety administrative system in Japan consists of related organizations and organizations responsible for food safety administration based on the legal system, their administrative duties, and related regulations. In addition to discussing the legal system related to food safety, the relevant organizations and organizations responsible for food safety administration, their administrative duties, and related regulations will be summarized.

The first is the risk assessment organization. The Food Safety Commission, established in the Cabinet Office in July 2003, has been appointed by the Prime Minister with the consent of both the House of Representatives and the House of Councilors,

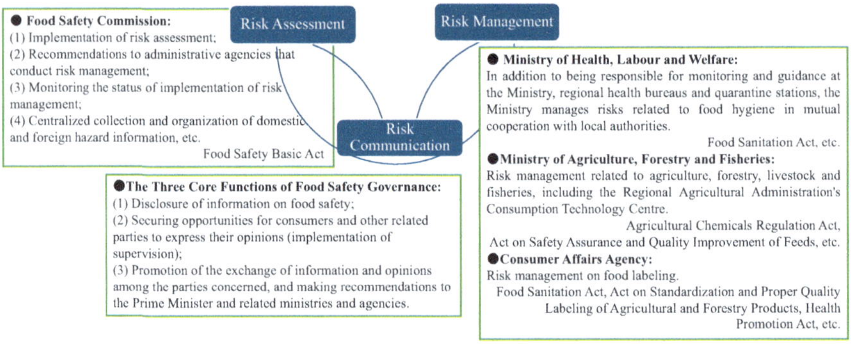

Fig. 5.4 Risk analysis initiatives in Japan. *Source* Expanded and created by the author based on the Ministry of Health, Labour and Welfare, Pharmaceuticals and Food Safety Bureau, Food Safety Department "Initiatives for Ensuring Food Safety" and information published on the websites of related ministries (Ministry of Health, Labour and Welfare; Ministry of Agriculture, Forestry and Fisheries)

and consists of 7 members with specialized knowledge (three-year term, reappointment possible; three of them are part-time). The Secretariat consists of the Secretary General, Deputy Director, General Affairs Division, Evaluation Division 1, Evaluation Division 2, Information and Recommendation Public Relations Division, Risk Communication Officer, and Evaluation Information Analyst. Based on the basic recognition that the protection of public health is of the utmost importance, the Food Safety Commission is an organization that conducts objective, neutral, and fair risk assessments based on scientific knowledge, independent of administrative agencies that manage risks such as regulations and guidance. Under this, 12 specialized committees have been established for each hazard factor, such as additives, pesticides, and microorganisms, including a special committee on planning, etc., to conduct risk assessments in an objective, neutral and fair manner based on scientific knowledge. The main role of the Food Safety Commission is to assess the risk of food at the request of the management organization and its own judgment, make recommendations to the relevant organizations, hold public meetings (in principle, every Thursday), post it on the homepage, in order to share information and exchange opinions with related parties including consumers in cooperation with related organizations and local governments. In addition, in the event of an emergency, it is necessary to cooperate with other organizations and provide scientific knowledge and opinions on hazardous substances to the public in a prompt and easy-to-understand manner through the mass media, government public relations, and the Internet.

The second is the risk management organizations. The Ministry of Health and Welfare focuses on risk management of food hygiene, such as the establishment of standards for additives and pesticide residues, the supervision of food processing, distribution, and consumption, and the introduction and certification of HACCP. Specifically, the Pharmaceutical and Food Safety Bureau is responsible for the overall coordination of administrative work, the examination of food hygiene standards, food and beverage safety monitoring, and the implementation of the HACCP Act. The Food Sanitation Division of each regional health and welfare bureau is responsible for the collection and disclosure of food consumption information, proper labeling monitoring, and public safety education in the region. The Ministry of Agriculture, Forestry and Fisheries is mainly responsible for the safety supervision of production materials and agricultural products in the production, distribution, and consumption of agricultural products, and risk management of agricultural products, such as the introduction and certification of GAP, and the specific work is mainly undertaken by the Food Safety and Consumer Affairs Bureau. In addition, Plant Protection Stations, seaports, and airport branches and branch offices stationed in Yokohama, Nagoya, Kobe, Moji, and Naha carry out import quarantine, export quarantine in response to the requirements of other countries, and domestic quarantine to prevent the spread of important pests. The head office is located in Yokohama and has seven branches at seven airports, preventing the invasion of infectious diseases, promoting livestock farming, and improving public health. In addition, the Food Safety and Consumer Affairs Bureau of each regional agricultural bureau is responsible for receiving and disseminating information on food safety in the region, and monitoring the traceability system and proper labeling of food in accordance with the JAS Act.

The third is the risk communication organization. The Consumer Affairs Agency, which was established in September 2009 as an external bureau of the Cabinet Office, serves as a control tower and engine for consumer administration. Regarding food safety, it is responsible for collecting, analyzing, and disseminating related information, supervising and managing agricultural products and food labeling, providing safety guidance on food consumption, and educating the public about food safety. At the same time, the Cabinet Office has established a Consumer Affairs Committee as a third-party organization. Responsibilities related to food safety are mainly independent investigation and deliberation of food safety, recommendations to the prime minister and relevant ministries, supervision of risk management by relevant ministries and agencies, and presentation of safety to management and consumers. The Consumer Affairs Committee is positioned as a council of the Cabinet Office and consists of up to 10 members appointed by the Prime Minister, as well as temporary members and expert committee members as necessary.

Furthermore, as demonstrated in Fig. 5.5, the exchange of information, the outsourcing of operations, the commissioning of operations, supervision and advice are all active among organizations related to "risk analysis." Of particular note is the establishment of a consumer safety information department by the Consumer Affairs Agency and related ministries and agencies. In ordinary circumstances, a system has been put in place to facilitate close exchange and communication among the parties concerned, including the dissemination of information to consumers and matters relating to food safety, such as the sharing of a list of people. In the event of an emergency, the government is collaborating to prevent the occurrence or spread of consumer damage by implementing a swift initial response system.

(d) Efforts to build consumer trust

The consumer's perspective is indispensable for food safety and security. As mentioned above, building consumer trust in the food safety administration system is both an important content and a purpose. Therefore, it is thought that trust can be built in the process of sharing information through risk communication. From this perspective, examining the framework of Japan's food safety administrative system, there are considered to be the following four initiatives aimed at information sharing with consumers and building consumer trust.

The first is a shift in the food safety administrative system to a "consumer mindset." Food safety administration has changed from "improving and promoting public health" to "focusing on consumers for the purpose of protecting the health of the public through ensuring food safety." In other words, the food safety administrative system is working to strengthen the monitoring and inspection system and provide information to businesses and consumers throughout the entire process from production to consumption. Furthermore, as part of risk analysis, risk communication is a process that guides countermeasures and "risk management" through information disclosure and exchange of opinions. Therefore, emphasis is placed on the disclosure and exchange of information to consumers and the presentation of safety.

The second is the introduction and dissemination of a food traceability system. Food traceability is a system that enables food businesses to grasp the movement of

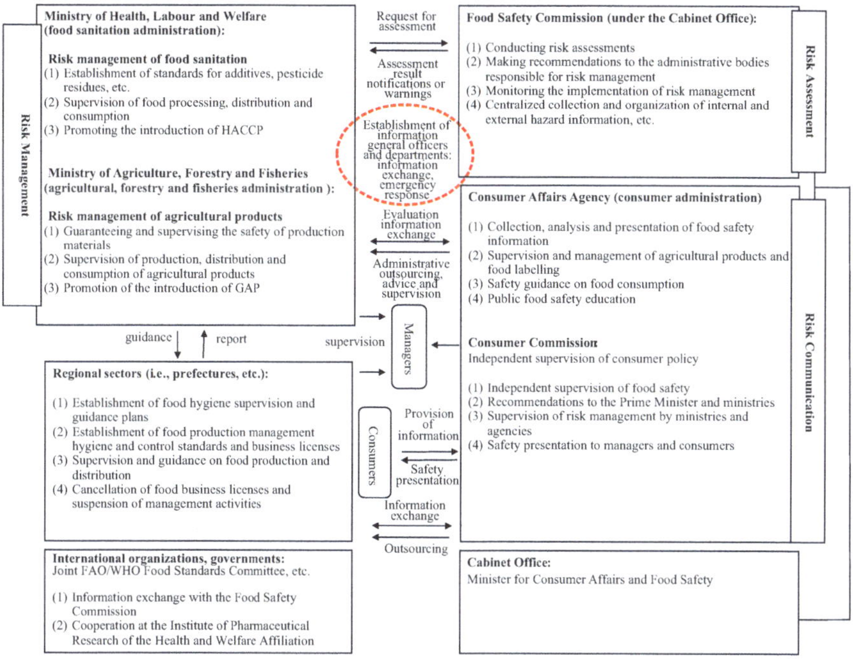

Fig. 5.5 Establishment of food safety administrative system in Japan. *Source* Expanded and created by the author based on Li & Nanseki (2013) and information published on the websites of related ministries (Ministry of Health, Labour and Welfare; Ministry of Agriculture, Forestry and Fisheries)

food from production to consumption by keeping records of food receiving and shipping destinations. There are two perspectives on the traceability system: "facilitating follow-up investigations and recalls in the event of a food accident" and "establishing a 'face-to-face relationship' with producers by providing production information to consumers and securing consumer trust." In particular, the latter is "traceability for security" through information disclosure, and can be positioned as a specific method of risk communication because it responds to consumers' questions about production and distrust of food safety through information disclosure. In Japan, traceability of cattle, rice and processed rice products is legally required.

The third is the proper regulation of food labeling. In September 2009, the Consumer Affairs Agency was established as a centralized consumer administration, and planning and enforcement of quality labeling standards under the JAS Act was transferred to the Agency. However, the Ministry of Agriculture, Forestry and Fisheries (MAFF) will continue to carry out on-site inspections, instructions to manufacturers and others to make improvements and food labeling monitoring. In order to ensure proper food labeling, MAFF's food labeling monitoring staff routinely monitor and provide guidance on labeling to retail outlets and collect information through the Food Labeling Hotline 110. In addition, at the regional level, food labeling monitoring councils are held with MAFF's outposts and relevant prefectural authorities as members, to share information and exchange opinions on inappropriate food labeling. Furthermore, the Food Labeling Act enacted on 28 June 2013 has further strengthened the standards and enforcement of food labeling.

The fourth is the promotion of food education. According to the Ministry of Agriculture, Forestry and Fisheries (2013), food education is the basis of life and should be the foundation of intellectual education, moral education, and physical education. In order to promote the practice of a Japanese-style diet, support is provided for wide-area, advanced nutrition education activities and nutrition education activities tailored to local conditions. Furthermore, useful information for a safe and healthy diet is disseminated, including correct knowledge on food safety and how to select and handle food appropriately. This is achieved through opinion exchange meetings and on the website. Furthermore, as part of food safety education for consumers, the company actively dispatches lecturers to local authorities, holds junior food safety seminars for junior high school students, and holds round-table discussions with the mass media and consumers from time to time to exchange opinions and provide information. Specifically, the Consumer Affairs Agency is collaborating with local governments and consumer groups to share information and exchange opinions with consumers, businesses, and experts to promote understanding of risks and safety that are familiar to consumers. The Ministry of Health, Labour and Welfare actively participates in opinion exchange meetings and roundtable meetings held by local governments and consumer groups, and strives to exchange information and opinions. In addition, the Ministry of Agriculture, Forestry and Fisheries is working to proactively provide information on topics related to ensuring the safety of foods, such as radioactive substances in food, and the prevention of animals and plants, through roundtable meetings with consumers and briefings for consumers and businesses in order to promote agriculture, forestry and fisheries administration from the consumer's point of view. From the perspective of risk communication, food education is of special significance for building consumer trust in food safety because it can provide scientific and objective information on food safety and promote interaction with consumers.

As the latest development since then, the Government of Japan reports an annual White Paper on Food Education to the Diet based on the Basic Law on Food Education, and the 2024 White Paper on Food Education features initiatives to deepen understanding of the agriculture, forestry and fisheries industries through hands-on experience in agriculture, forestry and fishing, and the promotion of "adult food education" for each generation as a special theme. These initiatives aim to increase the knowledge and behaviour of consumers themselves in relation to food, through cooperation between local communities, schools and private businesses. In particular, a wide range of measures are introduced for different generations and positions, such as promoting the public's participation in agriculture, forestry and fishing experiences, and spreading nutrition education activities to food business operators and employees. It also reports on the latest practical examples, such as the use of organic agricultural products in school lunches and specific examples of guidance on nutrition education, while examining the progress made in achieving the goals of the

Fourth Basic Plan for the Promotion of Nutrition Education. These developments are aimed at further spreading nutrition education and leading to the implementation of a healthy diet among the public.

5.2.2 *Laws and Administrative Systems for Ensuring Food Safety in China*

5.2.2.1 Structure of the Legal Framework for Ensuring Food Safety

The Chinese government enacted the Food Hygiene Law (Trial) in 1981 and began to manage food sanitation, but it was not until 1995, more than 10 years later, that the Food Hygiene Law was officially enacted. Subsequently, laws and regulations related to food quality inspection, market entry, etc. were rapidly developed. In addition, the Standardization Law (1989), the Product Quality Law (1993), the Implementation Rules for the Supervision and Administration on the Quality Safety of the Food Manufacturing and Processing Enterprise (2003), and the Measures for the Supervision and Administration of Food Safety in the Circulation Links (2006) for standardization including all stages of production, processing, distribution, and sales, as well as the Border Health and Quarantine Law (1987) and the Law on Import and Export Commodity Inspection (1989) regarding the quarantine of imported and exported agricultural products and foods, the Law on the Entry and Exit Animal and Plant Quarantine (1992) and the Animal Epidemic Prevention Law (1998) were enacted. However, due to the decline in morale of food manufacturers and problems with the government's monitoring and control system, the frequent occurrence of food safety problems has increased consumer interest and anxiety about food safety. At the same time, government leaders were keenly aware of the importance of food safety issues (Bian, 2012; Wu et al., 2018).

In particular, a serious food safety incident in 2008, when the chemical melamine was found in powdered milk produced by Chinese dairy manufacturer Sanlu Group, caused widespread damage to the health of infants. The melamine contamination caused many infants to develop kidney stones and other diseases, and several deaths were reported, revealing the serious impact on consumer health. The incident was widely reported in the mass media both at home and abroad, and triggered an increase in consumer distrust of Chinese dairy products and food products in general, and also dealt a serious blow to confidence in the Chinese government's food safety management system itself.

Against the backdrop of this incident, China replaced the previous "Food Hygiene Law" with a new law featuring more comprehensive and stringent provisions. On February 28, 2009, the Standing Committee of the National People's Congress adopted the 'Food Safety Law of the People's Republic of China,' which came into effect on June 1 of the same year. This law establishes a systematic framework for ensuring food safety, aims to protect public life and health, and clarifies safety standards and the regulatory framework for the production, processing, distribution, and sale of food.

Moreover, for agricultural products used as food ingredients, the Law of the People's Republic of China on Quality and Safety of Agricultural Products, which was enacted in 2006 and came into force the same year, continues to apply. This also regulates the quality and safety management of edible agricultural products by law and, together with the Food Safety Law, forms the basis of China's food safety system.

The legal system continued to evolve after the Food Safety Law came into force, and in 2015, the Law was substantially amended to introduce provisions described as "the strictest in history", including stricter penalties for illegal acts and measures to strengthen the management system.

In this way, the Sanlu Milk case has aroused attention to food safety not only in China but also internationally, and was a major turning point in the subsequent development and tightening of legislation and regulations. The Chinese Government has continued its efforts to develop a food risk management and supervision system, centered on the Food Safety Law, and the institutional infrastructure for consumer protection and ensuring food safety has been strengthened. The main legislation is shown in Table 5.4.

Three features of the evolution of China's legislation on ensuring food safety can be identified.

First, the focus has shifted from "physical health and fitness" to "life safety." As for the name of the law, it was revised from "Food Hygiene Law" to "Food Safety Law", same as in Japan. Prior to the enactment of the Food Safety Law, the Food Hygiene Law existed as the basis of food legislation. Article 1 of the "Food Hygiene Law" stated, "This Law is enacted for the purpose of ensuring food hygiene, preventing food contamination and harmful substances from injuring human health, safeguarding the health of the people and improving their physical fitness." On the other hand, the Food Safety Law states, "This Law is developed to ensure food safety and protect the physical health and life safety of the public." Comparing these two expressions, the purpose of the "Food Hygiene Law" was limited to physical health and fitness, while the "Food Safety Law" can be understood as a concept that leads to a broader "life safety" including the purpose of the "Food Hygiene Law." The transition from the "Food Hygiene Law" to the "Food Safety Law" signifies a shift in focus from mere physical health to a more comprehensive understanding of life itself. Based on the tragic lessons in which lives were lost, this shift implies the core principle of "Yi Ren Wei Ben" in Chinese or "people oriented" in English, positioning food safety not merely as an economic issue but as a public policy aimed at protecting the life and health of the people (Hu, 2007).

Second, the objects and scope of laws related to food safety have broadened. In particular, the Food Safety Law consists of 10 chapters and 104 articles, including the responsibilities of each entity related to food safety, food safety risk monitoring and assessment, food safety standards, responsibility for food production, distribution, and sales, food inspection, food import and export, food safety prevention and accident handling, and punishment provisions. Furthermore, while food safety

Table 5.4 Major laws and regulations related to food safety in China

Classification	Name of the law	Year of enactment and revision	Short description
General laws related to food safety	(1) Food Hygiene Law	Enacted in 1995, abolished in 2009	Regulations for monitoring and management of food hygiene
	(2) Food Safety Law	Enacted in 2009, revised in 2015	Strengthen monitoring and management, establish a new Food Safety Committee, establish a system for transparency of administration and information, establish a monitoring and risk assessment system, strengthen safety standards and market access systems, import and export food safety and penalties, prohibit excessive advertising, and begin to emphasize risk communication
Laws related to agricultural production	(3) Agricultural Law	Enacted in 1993, revised in 2003 and 2013	Establishment of quality standards and inspection and supervision systems for agricultural products. Pest and disease quarantine system, registration, permit, and safe use of pesticides and agricultural production materials
	(4) Agricultural Product Quality and Safety Law	Enacted in 2006	Agricultural products are in scope. The competent departments of agriculture at county level and above are responsible for the supervision affairs of the safety quality of agricultural products. The Ministry of Agriculture establishes a special committee for risk assessment of agricultural product quality and safety and carries out risk assessment and analysis. Promotion of agricultural product quality and safety standards, production area, production, packaging and labelling management, inspection supervision and penalties
	(5) Environmental Protection Law	Enacted in 1989, passed the draft amendment in 2014, came into effect in 2015	Protection of the environment related to agriculture and fisheries, such as soil and water resources, rational application of chemical fertilizers and pesticides, and safety of agricultural products

(continued)

Table 5.4 (continued)

Classification	Name of the law	Year of enactment and revision	Short description
Laws related to food production	(6) Product Quality Law	Enacted in 1993, revised in 2000 and 2009	Supervision of product quality, responsibility and obligation of producers and sellers for product quality, compensation for damages, and penalty provisions. Processed products are within the regulations
	(7) Standardization Law	Established in 1989, in revision	In view of the demand for uniform technology, the establishment of standards, implementation, and legal responsibilities are stipulated. The target is mainly industrial products, and only important agricultural products are targeted
	(8) Implementation Rules for the Supervision and Administration on the Quality Safety of the Food Manufacturing and Processing Enterprise	Established in 2003	Implementation of monitoring and management of food production stages, and provision of penalties
Laws related to the distribution and import of food products	(9) Measures for the Supervision and Administration of Food Safety in the Circulation Links	Established in 2006, revised in 2009	Implementation of monitoring and management of food distribution and sales, and provision of penalties
	(10) Law on Import and Export Commodity Inspection	Established in 1989, revised in 2002	Provision for quarantine of goods included in the import and export commodity inventory
	(11) Border Health and Quarantine Law	Established in 1987, revised in 2007	Goods such as immigration personnel, transportation equipment, luggage, cargo, and mail are eligible
	(12) Law on the Entry and Exit Animal and Plant Quarantine	Established in 1992, revised in 2009	Regulations on quarantine for import and export animals, animal and plant products, loading containers, and packaging

(continued)

Table 5.4 (continued)

Classification	Name of the law	Year of enactment and revision	Short description
	(13) Animal Epidemic Prevention Law	Established in 1998, revised in 2008	Prevention, control, eradication, reporting, and publication of animal diseases, quarantine of animals and animal products, animal health regulations are carried out in the process of animal breeding, rearing, slaughter, management, isolation, and transportation, and penalty provisions
Laws related to food labeling	(6) Production Quality Law	Established in 1993, revised in 2000 and 2009	Implementation of monitoring and management of mass authentication, production area, manufacturer, etc., and penalty provisions
	(7) Standardization Law	Established in 1989, in revision	Regulate the design, production, packaging, storage, use, etc. of products. The target is mainly industrial products, and only important agricultural products are targeted
Laws related to food consumption	(14) Law on the Protection of Consumer Rights and Interests	Established in 1994, revised in 2009 and 2013	Consumer rights, obligations of businesses, protection of legitimate rights and interests of consumers, consumer organization, dispute resolution, legal liability

Note The Food Hygiene Law was abolished at the same time as the Food Safety Law was enacted
Source Expanded and created based on publicly available materials

standards have been strengthened in the Food Hygiene Law, several features of the legislative policy and substantive provisions can be interpreted as follows. These include the establishment of a food safety risk assessment system, the abolition of the "inspection exemption system" for food commodities and the increase of compensation for damages (ten times). Nevertheless, there are still outstanding issues to be addressed, including the illegal use of additives, the inaccurate labelling of food products and the emergence of safety concerns related to imported food commodities, and the following characteristics can be read from legislative policies and substantive provisions.

Third, the scope of application to new food safety issues are expanded. In order to adapt to the new trends in the food safety situation from 2013, the "Food Safety Law" was revised, and the amendment was passed by the National People's Congress by April 2015. Even though the revision of the Law further tightened the contents related to online shopping of food products, genetic modification, health foods, etc., clarification of safety responsibilities, and the establishment of a monitoring and management system, there are still challenges in ensuring the safety of edible agricultural products. As shown in Table 5.5, quality and safety management at the production stage of edible agricultural products is mainly to comply with the provisions of the Agricultural Product Quality and Safety Law, while the relevant provisions of the Food Safety Law must be complied with when establishing quality safety standards for edible agricultural products and publishing safety-related information. Therefore, due to the diversification of laws related to ensuring the safety of edible agricultural products, it is considered that dual administration and unclear authority and responsibility in the administrative side still remain to adapt to new food safety situations.

Specifically, Article 2 of the Food Safety Law and Article 3 of the Agricultural Product Quality and Safety Law state that the agricultural sector of the government at the prefectural level and above is responsible for monitoring and controlling the quality and safety of agricultural products. However, Articles 33 and 52 of the Agricultural Product Quality and Safety Law stipulate that the industrial and commercial management department must monitor and control the amount of pesticide residues, heavy metals, and other substances that pose a danger to human health to agricultural products in agricultural products markets. This overlapping of functions and duplication of functions tends to lead to unclear authority and responsibility. The second is Article 29 of the Food Safety Law, which states that when a farmer sells his own edible agricultural products, it is not necessary to obtain a permit for food distribution. Article 77 of the Law states that the competent sectors of agriculture should comply with the provisions of the Agricultural Product Quality and Safety Law that monitors and controls edible agricultural products. However, the Agricultural Product Quality and Safety Law does not stipulate a system for granting permits for the sale and distribution of edible agricultural products. In other words, it becomes a blind spot in food safety monitoring and management measures at the stage of sales and distribution of edible agricultural products.

5.2.2.2 Current Status of the Food Safety Administrative System

The monitoring and control system for food safety in China can be broadly divided into two types of control: general administrative licensing control and technical control (standards and inspection). The first is the general administrative kind of control, which is mainly expressed in administrative permits for various food production, processing, management and consumption (such as hygiene permits, production permits, business licenses, etc.). The other is the technical kind of control, which is mainly expressed in the establishment of food standards and food inspection. There are few overlapping government bodies for these two types of control functions, which are basically jointly controlled by several single departments and committees, such as the Health Administration, the Ministry of Agriculture and Rural Areas, the State Administration for Market Supervision and Administration [formerly the State

Table 5.5 Application scope of food safety-related laws in China

		Production stage	Processing stage	Distribution, import/ export stage	Consumption stage
Food commodity	Processed food products	(2): Various provisions including provisions on quantitative limits for substances hazardous to human health; safety standards for food additives; provisions and management systems for risk assessment and risk management; provisions on illegal penalties.	(6) (7): Obligations of food producers to product quality, standards, and packaging, compensation for damages, and penalties. (8): Implementation of monitoring and management of food processing stages, and provision of penalties. *Implementing organization*: Center for Food and Drug Inspection, related departments such as sanitation, industry and commerce, quality inspection, etc.	(2): Safety standards, norms, and penalties for food distribution, sales, and food service activities. (9): Implementation of monitoring and management of food distribution and sales, and provision of penalties (10) (11) (12) (13): Safety standards for import and export food, methods of inspection and quarantine, penalty provisions, etc. *Implementing organization*: Center for Food and Drug Inspection, related departments such as sanitation, industry and commerce, quality inspection, etc.	(2): Dissemination of food safety laws, standards, and knowledge, and provisions to improve consumers' safety awareness and self-protection capabilities. (14): Provisions on consumer rights, obligations of businesses, protection of legitimate rights and interests of consumers, consumer organization, dispute resolution, and legal liability. *Implementing organization*: China Consumers Association
	Edible agricultural products	(3): Establishment of quality, standards, and supervision systems for agricultural products; (4): The production of edible agricultural products is generally stipulated from safety standards to packaging and labeling at the time of shipment. *Implementing organization*: Ministry of Agriculture and agricultural departments at the prefectural level and above			

Note 1 The numbers in () are the numbering of the laws in Table 5.4
Note 2 The implementing organization at each stage refers to the content of the later part of the chapter on the administrative structure
Source Created based on Table 5.4 and related materials

Administration for Quality Supervision, Inspection and Quarantine] and the State Administration for Industry and Commerce. Some commentators have stated that it is difficult to manage one pig with eight different departments and committees. Statistics show that the monitoring and control systems related to food safety in the central authorities such as health, agriculture, quality inspection, industry and commerce, as well as in the local administrative bodies and business units affiliated to them, are huge and various personnel, including those of the former State Food and Drug Administration (CFDA), have been assigned to form a vast monitoring and control system, yet the effectiveness of this system is not always clear. Despite this, some point out that the effectiveness of the system is not always consistent and clear.

In response to this background, the enactment of the Food Safety Law has led to the establishment of a mechanism to reiterate and clearly define the division of responsibilities among the relevant administrative structures and departments for supervision and management of food safety, and to strengthen cooperation among relevant ministries and agencies. The Law comprehensively and in detail defines norms and standards across all stages of food production, marketing and other activities, and strengthens the unified coordination of relevant administrative bodies in the assessment and management of food safety risks and the treatment of food safety incidents.

Furthermore, the latest amendment to the Food Safety Act passed in 2025 includes provisions to strengthen safety supervision throughout the supply chain, such as ensuring the quality and safety of liquid infant formula and introducing a road transport permit system for bulk liquid food, which is characterized by a clearer responsibility relationship between companies and transporters. This is expected to lead to more segmented safety management and stricter penalties for non-compliance than in the past. Furthermore, in order to strengthen comprehensive national food safety supervision and management, the State Council's guidelines will be issued in March 2025, proposing measures to improve the "farm to table" all-process supervision system, strengthen supervision coordination of school meals and food delivery services, and review inspection and quarantine certification procedures.

On the technical management side, the State Administration for Market Supervision and Administration and the State Health and Health Commission are taking the lead in revising and establishing new national food safety standards (GB standards), with several new standards (e.g. limits of contaminants in food, labelling and additive standards) being promulgated in 2025. Progress has also been made in the monitoring and inspection function in practice, with sampling inspections by the State Administration of Market Supervision and Administration routinely conducting food safety inspections across the country, and a decrease in the failure rate has been observed.

Thus, China's food safety administration system is undergoing a shift from the traditional cross-agency management system to a more integrated and strengthened cooperation through clarification of the division of responsibilities among relevant agencies and revision of laws and regulations, which is improving the efficiency and effectiveness of the food safety supervision and management function.

In addition, in China, the food safety monitoring and management system can generally be divided into two types: administrative control through permits and technical management (standards and testing). The first type pertains to general administrative oversight, mainly manifested in various administrative permits related to food production, processing, operation, and consumption (such as hygiene permits, production permits, and business licenses). The second type involves technical management, primarily reflected in the formulation of food standards and food testing. Although there has often been overlaps among the government agencies responsible for these two types of management functions, administrative reforms and departmental integrations have been making the system more organized.

At the first session of the 12th National People's Congress on March 14, 2013, the departmental organization of the State Council (equivalent to the Central Government Ministries and Agencies of Japan) was established. The number of departments was reduced from 27 to 25, excluding the General Office of the State Council. By merging the administrative departments related to food safety, the National Health Commission (formerly the National Health and Family Planning Commission, which dissolved the Ministry of Health, and National Population and Family Planning Commission) and the China Food and Drug Administration (which dissolved the State Food and Drug Administration, affiliated with the former Ministry of Health) were established.

The National Health Commission integrates the functions of the former Ministry of Health (including disease prevention, medical systems, and health policy) and the National Population and Family Planning Commission, overseeing public health, disease control, health policy, and population management in a unified manner.

The National Medical Products Administration is a newly established agency that consolidates the food safety-related teams formerly under the State Administration for Industry and Commerce and the food and pharmaceutical regulatory functions that belonged to the Ministry of Health (formerly the Food and Drug Administration). It has been reorganized to comprehensively oversee the production, distribution, and consumption stages of food and pharmaceuticals.

Since then, progress has been made in further organizing and strengthening the organizational structure for food safety: since the amendment of the new Food Safety Law in 2018, a system has been established in which the State Administration for Market Supervision and Administration (SAMR) functions as the central food safety supervision body. This is the result of the integration of the former State Administration of Quality Supervision, Inspection and Quarantine (AQSIQ), State Administration for Industry and Commerce (SAIC) and the State Administration for Food and Drug Administration (CFDA), which centralized food safety-related supervision and enforcement functions. The State Administration for Market Supervision is responsible for food safety licensing, inspection and quarantine, labelling supervision and violation handling from the raw material stage to the consumption stage in accordance with the Food Safety Law.

On the technical management side, led by the State Administration for Market Regulation and the National Health Commission, national food safety standards (GB standards) are being revised and newly established, with a number of new standards (e.g., limits for contaminants in food, labeling, and additive standards) promulgated in 2025. In addition, the State Administration for Market Supervision is developing a routine safety monitoring system through sampling inspections and publication of rejected products in markets across the country.

Furthermore, the latest amendments to the Food Safety Law in 2025 include new provisions such as strengthening quality control of liquid infant formula, introducing a road transport management system, clarifying the food safety risk assessment system and increasing corporate responsibility. The amendments aim to ensure safety throughout the supply chain, improve the efficiency of administrative supervision and clarify sanctions in the event of non-compliance. In addition, the joint Party Central and State Council food safety guidelines (March 2025) set out a policy to strengthen the "farm-to-table" supervision system and the response to new consumption patterns such as school meals and online sales.

Thus, China's food safety administration system is undergoing a shift from the former decentralized system with many overlaps between departments to a unified supervision and management system based on the integration of functions and strengthening of coordination by the center. In particular, the consolidation of functions centered on the State Administration for Market Supervision and Administration is being operated as a mechanism covering all stages of licensing, standard setting, inspection and violation handling, and a more efficient and clear responsibility system is being developed than before (Bian, 2012).

Figure 5.6 shows the current food safety administration system in China.

The first is the Food Safety Committee of the State Council. In accordance with the Food Safety Law, the Food Safety Committee was established by the State Council in February 2010 as the highest guiding and coordinating body for the food safety management system. The then Deputy Prime Minister was appointed as the head of the committee, two deputy chiefs were appointed, and 15 members were appointed as committee members, including heads and directors of related departments such as sanitation, agriculture, quality inspection, and industry and commerce. The committee is primarily responsible for the overall coordination of food safety operations: (1) analysis of the current state of food safety, assignment of work and coordination of functions between departments to ensure food safety; (2) submission of major policy measures on food safety supervision; and (3) supervision of compliance with food safety regulations. In addition, according to the organizational reform of the State Council in March 2013, the China Food and Drug Administration established the Food Safety Commission Secretariat, and the daily affairs of the Food Safety Commission of the State Council were transferred to the China Food and Drug Administration. In April 2014, the first meeting of the expert committee of the Food Safety Commission of the State Council, consisting of 51 experts in various fields related to food safety, was established. The expert committee is mainly responsible for technical consultation on food safety, advice on policymaking, science dissemination and promotion, and is managed by the Office of the Food Safety Commission of the State Council. The Food Safety Commission of the State Council is tasked with providing guidance on food safety on a day-to-day basis, but its chief, deputy director, and members each have their own full-time duties, so it is believed that it

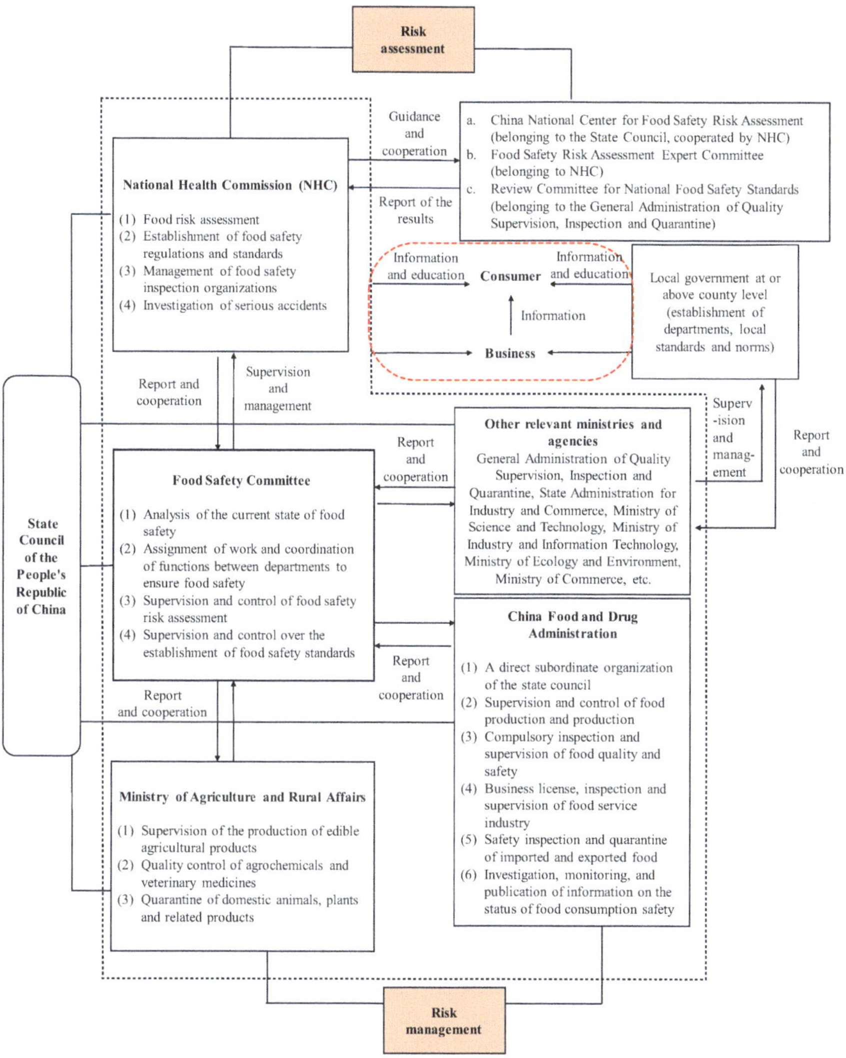

Fig. 5.6 Current status of food safety administration system in China. *Note* The State Council Food Safety Commission serves as the highest leadership and coordination body for food safety management, guiding and supervising the food safety operations of the National Health Commission, the Ministry of Agriculture, and other relevant ministries. *Source* Expanded and created by the author based on sources such as the State Council of the People's Republic of China (2019)

may become an agency that only conducts temporary proceedings and coordination and directs major accidents and incidents.

The second is the National Health and Family Planning Commission (which was abolished following the Chinese Government's structural reform in March 2018, with many of its functions being taken over by the newly established National Health

Commission). As mentioned above, this commission inherited the functions of the former Ministry of Health and is involved in formulating and supervising the implementation of basic public health, disease control and food safety strategies. In the area of food safety, it plays an important role, including risk assessment, formulation of science-based standards and policies, coordination of the overall food safety management system, health impact assessment and epidemiological studies, with the following functions: (1) overseeing and promoting food risk assessment, (2) supporting the formulation of food safety regulations and standards, (3) food safety inspection and support for the evaluation and improvement of food safety inspection and monitoring systems, and (4) support for the investigation of serious food safety incidents and health hazard cases and investigation into the causes of such incidents.

As a food risk assessment mechanism, the National Centre for Food Safety Risk Assessment (NCFSA), building on the existing structure, plays a central role in risk assessment in cooperation with the National Health and Health Commission and the State Administration for Market Regulation (SAMR), plays a central role in risk assessment. In scientific assessment, the Expert Committee on Food Safety Risk Assessment is affiliated to the NHC and plays a role in reflecting expert opinion and scientific findings in policy making. In addition, the National Food Safety Standards Review Committee operates under the State Administration for Market Supervision and Administration and carries out review and coordination on the development and revision of the National Food Safety Standards (GB Standards).

In the latest developments, the linkage between risk assessment and standard-setting has been clearly established as an institutional system in line with the strengthening of the food safety system from 2021 onwards. In addition, the development and revision of national food safety standards has accelerated, with key national standards such as Limits of Contaminants in Food (GB 2762) and Standards for the Use of Food Additives (GB 2760) being updated to 2025 standards. This has further strengthened the science-based monitoring and evaluation system for food safety. Furthermore, the National Health and Sanitation Commission, in conjunction with the NHC's Public Health Monitoring Network, has developed a national food safety risk information collection and sharing system to improve food accident prevention and early response capacity.

Based on the assessment results of these various risk assessment bodies, the National Health and Health Commission supports the formulation of comprehensive food safety strategies and national food safety plans, while at the same time promoting the improvement and implementation of food safety management systems in coordination with the State Administration for Market Supervision and Administration. In particular, guidelines have been developed to improve the policy use of risk assessment results and the transparency of the standards revision process, and the formation of a unified food safety policy based on scientific evidence is being strengthened (Wu et al., 2018).

Third, the China Food and Drug Administration is a direct subordinate organization of the State Council, and as one of the responsible entities for food safety risk management, its functions are (1) supervision and control of food production and production; (2) compulsory inspection and supervision of food quality and safety;

(3) business license, inspection and supervision of food service industry; (4) safety inspection and quarantine of imported and exported food; and (5) investigation, monitoring, and publication of information on the status of food consumption safety. In other words, risk management is carried out for processed foods, food and beverage service industries, and import/export foods based on the results of risk assessments. In addition, the Administration has established the Secretariat of the Food Safety Commission, which also serves as a function.

The fourth is the Ministry of Agriculture (now known as the Ministry of Agriculture and Rural Affairs, MARA). The MARA is a constituent department of the State Council that is responsible for the economic development of agriculture and rural areas. The work of the Ministry of Agriculture and Rural Affairs in the area of food safety monitoring and management is mainly responsible for monitoring and managing the production stage of edible agricultural products as risk management, and following food safety-related affairs: (1) drafting of laws and regulations related to agriculture and fisheries and related industries, (2) quality supervision of agricultural and fishery industries and related industrial products and green foods, (3) quality monitoring and appraisal of pesticides and veterinary drugs registration; (4) supervision of domestic animal and plant protection and quarantine operations, etc.

In addition, the following relevant ministries and agencies each play a role in food safety administration in China: the State Administration for Market Regulation (SAMR), overseeing quality and safety of food production and processing enterprises, as well as import/export food inspection and quarantine; the Ministry of Science and Technology (MOST), conducting research and development related to food safety management; the Ministry of Industry and Information Technology (MIIT), managing networking and sharing of information related to food safety; the Ministry of Ecology and Environment (MEE), protecting the environment of agricultural production areas; the Ministry of Commerce (MOFCOM), supervising and managing market distribution and transportation of food; and the Ministry of Public Security (MPS), investigating crimes related to food safety. Moreover, local governments below the provincial and autonomous regions (at the prefectural level and above) also have local administrative departments that are in charge of food safety, such as sanitation, quality inspection, industry and commerce, similar to those of the national government, and together with the administrative agencies at the national level, establishing China's current food safety supervision system.

5.2.3 China's Food Safety Legal and Administrative Systems from a Comparison of Japan and China

5.2.3.1 Challenges in China

Japan and China are very similar in terms of agricultural types, staple foods and eating habits of the people. Although the national circumstances and social backgrounds of the two countries are different, the experience of Japan's legal and administrative systems as a developed country is an important reference when considering China's future economic development, environmental protection, and ensuring social security, especially food safety (Bian, 2012; Wu et al., 2018). First, Tables 5.6 and 5.7 compare the status of food safety legislation in Japan and China, focusing on the field and content of law.

As shown in Tables 5.6 and 5.7, the following points can be identified as experiences with Japanese legal and administrative systems regarding food safety, as well as institutional problems and issues related to ensuring food safety in China.

(a) Legal system

A review of food safety laws and administrative systems in China, as well as their implementation and management systems, reveals at least the following two problems. The first is that while Japan has detailed legislation to deal with problems in each field, in the case of China, laws and regulations related to ensuring food safety have been enacted one after another since the 1980s, but the year of enactment is late, and there are no laws in diversified fields like in Japan. In addition, it is often expressed as a "regulation" rather than a "law" (in China, a "regulation" is an administrative regulation established by the State Council, the administrative organ, on the basis of the Constitution and laws, which is subordinate in its effectiveness to the law).

Second, while laws relating to food safety in Japan are amended in a timely and appropriate manner, laws in China are not always amended quickly and many laws are often in operation without amendment for a long time after promulgation (see Table 5.8).

Laws relating to food safety and soil contamination in Japan have been amended in a timely and appropriate manner, and their operation is reported in detail. For example, the Soil Contamination Countermeasures Act has been amended four times since its enactment in 2002 until June 2011, with the aim of ascertaining the state of contamination of soil with specified hazardous substances and preventing health hazards. These amendments have expanded the system for ascertaining the state of contamination, clarified measures based on the classification of regulated areas and introduced new provisions to ensure the proper treatment of soil contamination. In 2023, the number of soil contamination survey reports reached 1,509, and the total number of areas designated as requiring action reached 6,938. 2021 saw a partial revision of the relevant standards, clarifying survey procedures when land use changes. Continuous improvement of the system is ongoing, with the Environment Council currently discussing a review of the law to address new contaminants such as PFAS. As for food safety laws and regulations, the provisions of the Food Sanitation Act and the Food Labelling Act are being updated between 2024 and 2025, and the development of science-based standards is steadily progressing.

Meanwhile, in China, the National Soil Contamination Situation Survey Bulletin published in 2014 showed that the total soil contamination rate nationwide reached 16.1% and the contamination rate of arable land 19.4%, indicating that more than 10% of arable land area is contaminated by heavy metals and other substances.

Table 5.6 Comparison of major laws related to food safety between Japan and China

	Japan	Year	China	Year
General laws related to food safety	Food Sanitation Act	1947 (2014)	Food Safety Law	2009 (2015)
	Food Safety Basic Act	2003 (2014)		
Laws related to agricultural production and food production	Livestock Hygiene Service Centers Act	1950 (1999)		
	Act on the Prevention of Infectious Diseases in Livestock	1951 (2014)	Animal Epidemic Prevention Law	1998 (2008)
	Act to Prevent Soil Contamination on Agricultural Land	1970 (2014)	No specialized law (Environmental Protection Law)	1989 (2015)
	Act on Stabilization of Supply, Demand and Prices of Staple Food	1995 (2011)		
	Basic Law on Food, Agriculture and Rural Areas	1999 (2009)	No specialized law (Agricultural Law)	1993 (2012)
	Act on Promotion of Introduction of Sustainable Agricultural Production Practices	1999 (2013)	No specialized law (Agricultural Law)	1993 (2012)
	Sustainable Aquaculture Production Assurance Act	2003 (2014)		
	Act on the Establishment of Laws Related to the Ministry of Agriculture, Forestry and Fisheries to Ensure Food Safety	2003 (2013)		
	Act on the Conservation and Sustainable Use of Biological Diversity through Regulations on the Use of Living Modified Organisms	2003 (2014)	Regulations on Administration of Agricultural Genetically Modified Organisms Safety	2001
	Act on Promotion of Organic Agriculture	2006 (2011)	No specialized law (Agricultural Law)	1993 (2012)
	Act on the Quality Control of Fertilizer	1950 (2011)	No specialized law (Environmental Protection Law)	1989

(continued)

Table 5.6 (continued)

	Japan	Year	China	Year
	Act on Safety Assurance and Quality Improvement of Feeds	1953 (2007)	Regulation on the Management of Feeds and Feed Additives	1999
	Slaughterhouse Law	1953 (2014)	Regulations on the Administration of Slaughtering of Pigs	1997
	Licensed Cooks Act	1958 (2014)		
	Confectionery Hygiene Control Law	1966 (2014)	Product Quality Law	1993 (2000)
	Poultry Slaughtering Business Control and Poultry Meat Inspection Act	1990 (2014)		
	Act on Temporary Measures concerning Sophistication of Management of Food Manufacturing Process (or HACCP Act)	1998 (2013)		
	Act on Promotion of Recycling and Related Activities for Treatment of Cyclical Food Resources	2000 (2013)		
	Act on Special Measures concerning Measures against Bovine Spongiform Encephalopathy	2002 (2003)		
Laws related to the distribution and import of food products	Import Trade Control Order	1949 (2003)	Law on Import and Export Commodity Inspection	1989 (2002)
	Agricultural Products Inspection Act	1951 (2014)	Agricultural Product Quality and Safety Law	2006
	Quarantine Act	1951 (2014)	Law on the Entry and Exit Animal and Plant Quarantine	1991 ara>
	Plant Protection Act	1950 (2014)	Regulations on Plant Quarantine	1983

(continued)

Table 5.6 (continued)

	Japan	Year	China	Year
	Act on Special Measures concerning Prevention of Toxic Contamination of Food Distributed Through Marketing Channels	1987 (2013)	Measures for the Supervision and Administration of Food Safety in the Circulation Links	2006 (2009)
	Beef Traceability Act	2003 (2008)		
	Act on the Recording of Information Related to Transactions of Rice Grain and the Transmission of Information on Production Areas	2009 (2013)		
	Act on Standardization and Proper Quality Labeling of Agricultural and Forestry Products (or JAS Act)	1950 (2014)	No specialized law (Agricultural Product Quality and Safety Law)	2006
	Measurement method	1951 (2014)	Standardization Law	1989
	Act against Unjustifiable Premiums and Misleading Representations	1962 (2014)	No specialized law (Standardization Law)	1989
	Household Goods Quality Labeling Act	1962 (2011)	No specialized law (Standardization Law)	1989
	Act on the Promotion of Sorted Collection and Recycling of Containers and Packaging	1995 (2011)		
	Health Promotion Act	2002 (2014)		
	Food Labeling Act	2013 (2014)	No specialized law (Standardization Law)	1989
Laws related to the control of agricultural residues in food	Agricultural Chemicals Regulation Act	1948 (2014)	Regulation on Pesticide Administration	2009
	Act on Securing Quality, Efficacy and Safety of Products Including Pharmaceuticals and Medical Devices	1960 (2014)	Regulation on Veterinary Drug Administration	2004

(continued)

Table 5.6 (continued)

	Japan	Year	China	Year
Regulations related to radioactive material inspections of agricultural, livestock and fishery products	Restrictions on intake based on the Act on Special Measures for Responding to Nuclear Disasters	1999 (2014)		
	Concept of setting and canceling inspection plans, shipping restrictions, etc. (laws and regulations)	2011 (2014)		
	Inspection of radioactive substances in agricultural, livestock and fishery products (laws and regulations)	2013		
Import-related laws (animals, plants, and food-related)	Import Trade Control Order	1949 (2003)	Regulation on the Administration of the Import and Export of Goods	2002
	Act on the Protection of Marine Resources	1951 (2010)	Regulations on the Protection and Reproduction of Aquatic Resources	1972
	Act on the Prevention of Adverse Ecological Impacts Caused by Designated Invasive Alien Species	2004 (2014)	Law on the Entry and Exit Animal and Plant Quarantine	1992
			Border Health and Quarantine Law	1987
Others	School Lunch Program Act	1954 (2008)		
	Act on the Regulation of Manufacture and Evaluation of Chemical Substances	1973 (2014)		
	Product Liability Act	1994	Product Quality Law	1993 (2000)
	Enforcement Regulations of the Act on Special Measures against Dioxins	1999 (2010)		
	Basic Act on Consumer Policies	1968 (2012)	Law on the Protection of Consumer Rights and Interests	1994 (2013)
	Basic Act on Shokuiku (Food and Nutrition Education)	2005 (2009)		

(continued)

Table 5.6 (continued)

	Japan	Year	China	Year
Food safety general principles	Food Sanitation Act	(2018)	Food Safety Law	(2015)
Food additives and standards	Additive regulations under the Food Sanitation Act	Continuing amendment	National Standard for Food Safety (GB standard)	Continuing amendment
Food labeling	Food Labeling Act	2015	Food labelling regulations under the Food Safety Law	2015
Food import and export regulations	Regulation on the supervision and administration of imported foods under the Food Sanitation Act	Continuing amendment	Measures of the People's Republic of China for the Administration of Safety of Imported and Exported Food	2019
Food safety management systems	Food Sanitation Act, Food Safety Basic Act	2018	National Food Safety Law, Food Safety Supervision and Control of the National Health and Health Commission	2015
Food risk assessment	Risk assessment under the Food Safety Basic Act	2018	National Centre for Food Safety Risk Assessment (NCFSA) established	2015 established

Note 1 The "Year" is the time of enactment of the law, and the time of the most recent revision is shown in (). For the period up to May 2016, the dates of enactment and last amendment are given; from May 2016 onwards, for Japanese legislation, the focus is on the year of amendment; for Chinese legislation, the year of enactment/amendment and the year of establishment of the relevant body
Note 2 "No specialized law" means there is no directly related law
Source Compiled and created by the author

However, since the Soil Pollution Prevention and Control Law was enacted in 2018 and came into force in January 2019, basic provisions have been established for the investigation and prevention of soil pollution. Furthermore, in 2022, the Black Soil Protection Law was passed to protect black soil areas and prevent pollution, thereby systematically strengthening soil conservation in China's major grain-growing areas. However, while these laws and regulations have been developed, there have been few frequent amendments or comprehensive reviews of the laws and regulations, and implementation and supervision at the local level has varied. In the field of food safety in particular, progress has been made in the development of science-based standards, with the National Standard for Food Safety (GB Standard) being substantially revised in September 2025 and 32 standards, including the limit of contaminants in food, being renewed, but the development of a system to ensure

Table 5.7 Comparison of current administrative systems related to food safety in Japan and China

Items	Japan	China
Risk Analysis: (1) (2) (3)		
(1) Risk assessment	Led by the Food Safety Commission (established by the Cabinet Office); supervise the implementation of risk management, centralized collection and organization of hazard information	Led by the National Health Commission (a subsidiary of the State Council) and the Minor Evaluation Expert Committees and Centers; risk assessment, establishment of safety regulations and standards, management of food safety, inspection, and quarantine institutions, and investigation and investigation of serious accidents, etc
(2) Risk management	The Ministry of Health, Labour and Welfare is responsible for "food sanitation administration" and the Ministry of Agriculture, Forestry and Fisheries is responsible for "agriculture, forestry and fisheries administration"	The China Food and Drug Administration and the Ministry of Agriculture and Rural Affairs are in charge of the overall supervision and control of food production, distribution, import/export and consumption; the Ministry of Agriculture and Rual Affairs supervises and manages only the production stage of agricultural products for food
(3) Risk communication	The Consumer Affairs Agency and the Consumer Affairs Committee are responsible for information gathering, supervision, guidance, education, advocacy, exchange, etc.; the Consumer Affairs Agency is the control tower for the centralization of consumer administration; establishment of an information supervisor and department to exchange information and respond to emergencies	Risk communication aimed at improving trust in food safety is still in its infancy, and consumers do not have the ability to assess the situation or respond to misleading or misleading reports about food safety; no government agency to respond
Organizational structure of the administrative system	The Food Safety Commission leads risk assessments, supervises the implementation of risk management, and collects and organizes hazard information in a centralized manner; the Consumer Affairs Agency is the control tower for the centralization of consumer administration, and its main function related to food safety is to manage food safety information and appropriate labeling, as well as risk communication such as public guidance and education in cooperation with risk management agencies	The Food Safety Committee of the State Council is the control tower of the food safety administrative system, and the relevant ministries, departments, committees, and general bureaus work together under the direction of this control tower to fulfill their respective responsibilities; coordination with central agencies will be diminished, and their leadership and policy implementation will be less efficient

(continued)

Table 5.7 (continued)

Items	Japan	China
Consumer	The Consumer Affairs Agency, an administrative agency, has been established as a control tower for the centralization of consumer administration, and information processing related to accidents (Consumer Policy Division, Safety Division), and food education (Consumer Affairs Information Division) Emphasis is placed on food safety awareness and knowledge and information on laws and systems necessary for dietary habits; the Consumer Affairs Agency and the Consumer Affairs Commission collect, analyze, and announce food safety information, and provide information and present safety; related organizations and consumers directly cooperate with risk communication to collect information, advice, complaints, etc. from consumers through the National Consumer Affairs Center under the jurisdiction of the Regional Cooperation Division; work to ensure food safety; through direct and effective participation and transparency of the administrative system, consumers have a high level of trust in food safety administration	The Industrial and Commercial Administration Department shall establish specific measures and methods to protect the rights of consumers, investigate and dispose of illegal activities, and carry out consumer rights protection activities; consumer associations at all levels established under the Law on the Protection of Consumer Rights and Interests are social organizations with a quasi-governmental character, and carry out social supervision of food safety and protection of consumer rights; In addition, efforts to prevent consumer harm are extremely insufficient. There is no established food safety consumer administration, and consumers have low confidence in food safety and its management system. In particular, consumer awareness and awareness of food safety in rural areas and safety management of agricultural products are issues
Business operator	Risk management, risk communication, and local departments exchange information and supervise based on relative uniformity of regulations and standards	In addition to the risk management department, central and local departments and companies may be supervised by different standards and norms

Source Compiled and created by the author with reference to Wang et al. (2014, 2016), Liu and Zhang (2015), and National People's Congress of the People's Republic of China (2015)

consistent implementation of these systems across the country remains a challenge (Bian, 2012; Wu et al., 2018).

In this way, while in Japan laws are frequently amended to reflect social changes and scientific knowledge, with continuous strengthening of systems in areas such as soil contamination and food safety, in China, although the development of specialized laws addressing major environmental risks has progressed in recent years, many of these laws have long-term application after their promulgation, and comprehensive revisions are not carried out frequently. In recent years, the Chinese government has

Table 5.8 Japan vs China: key points of comparison of food safety and soil-related legal systems

Item	Japan	China
Soil contamination-related legislation	Soil Contamination Countermeasures Act (enacted in 2002), amended and revised several times after enactment, and environmental standards reviewed	Soil Pollution Prevention and Control Law (enacted in 2018 → entered into force 2019) and Black Soil Protection Law (2022) enactments have made progress, but tend to be amended less frequently due to institutional operations
Risk assessment and management standards	Ongoing review and reporting system for soil environmental standards, notification system for change of use, etc., established	Standards development after enactment of the Soil Pollution Prevention and Control Law has progressed, but local implementation and supervision systems are considered to be uneven across regions
Updated food safety standards	Food Sanitation Law and other related standards continuously revised and updated (food labelling, additives, etc.)	Progress in scientific testing and standards development due to the 2025 revision of large-scale national food safety standards (e.g. GB 2762–2025)
Frequency of amendments/ legislative adaptability	Frequent reviews to reflect social needs and scientific findings	Legislation is being developed, but is often in long-term operation after promulgation and the frequency of full revision is relatively low

strengthened the development of relevant standards and specialized laws, and attention is now focused on future amendments and improvements in their implementation by local governments.

(b) Administrative system

First, the Consumer Affairs Agency in Japan has independent supervisory authority, and as a control tower for the centralization of consumer administration, it supervises and advises the responsibilities of each ministry and agency. In addition, the organizational structure of the relevant ministries and agencies is under the direct control of the central government (vertical management), and they work together efficiently. On the other hand, although China clearly stipulated the responsibilities and division of duties of each relevant administrative agencies after the 2012 organizational reform, it is still questionable whether it is feasible to eliminate dual administration, ensure cross-sectional coordination among each agency, and provide reliable administrative services that businesses and consumers can understand. Next, in food safety risk management in Japan, the food sanitation administration led by the Ministry of Health, Labour and Welfare and the agriculture, forestry and fisheries administration led by the Ministry of Agriculture, Forestry and Fisheries ensure the safety of processed foods and agricultural products, respectively. In China, there are still challenges in ensuring the safety of edible agricultural products, as the laws on ensuring the safety of edible agricultural products are diversified, the overlap of supervisory departments, and the lack of clarity of authority and responsibilities are still apparent. As shown in Table 5.9, local departments related to food safety other than import/export inspections and quarantines have affiliated organizations at each level of local governments at the prefectural level and above, and these local agencies are strongly influenced by the local governments to which they belong. Some local governments have established their own food safety standards and ordinances, and the efficiency of their prompt guidance and policy implementation has been questioned by each central management body related to food safety. It also points out the cover-up of information about food safety incidents and accidents by some local governments. Finally, as mentioned above, China's Food Safety Committee, although tasked with guiding food safety on a day-to-day basis, has the potential to become a body that only conducts ad hoc proceedings and coordination and directs major accidents and incidents, and there are questions about how well it functions in practice.

(c) Consumer response

In Japan, the Consumer Affairs Agency, which is the administrative body of the country, has been established to process information on cases (Consumer Policy Division, Safety Division) and public safety education (Consumer Affairs Information Division), collect information, advice, complaints, etc. from consumers through the National Consumer Affairs Center under the jurisdiction of the Regional Cooperation Division, provide information and present safety to consumers, and work to ensure the safety and trust of food. In addition, in line with the promotion of food education as a national movement, policies are being implemented not only to promote healthy eating and drinking for consumers, but also to provide basic knowledge about food safety.

On the other hand, In China, however, risk communication on food safety (exchange and sharing of food safety information) is still in the process of development and an institutional framework has not been fully established. Currently, no specialized department or mechanism has been established to systematically promote information dissemination and dialogue with consumers on food safety, and there are limited mechanisms to systemize risk communication at the provincial level. Although relevant bodies such as the National Health Commission (NHC) and the State Administration for Market Regulation (SAMR) are responsible for developing food safety standards and supervision systems, institutionalized mechanisms focusing on risk information sharing and consumer participation are still limited.

Although industrial and commercial management departments and consumer associations at all levels have developed activities to protect consumer rights and deal with food safety damage, a comprehensive administrative system focusing on the prevention of consumer damage and food education is not fully in place. Specifically, while emphasis is placed on coordinating disputes between consumers and businesses and supporting the resolution of food damage, the system for raising consumer awareness and promoting preventive measures through education and provision of

Table 5.9 Organization and management of food safety supervisory agencies in China

Governing authority	Organizational structure	Management method	Supervisory & managerial capabilities
Ministry of Agriculture and Rural Affairs	Established at each level of local government at the prefectural level and above	Focusing on horizontal management	Supervision of the production of edible agricultural products
Health Commission	Established at each level of local government at the prefectural level and above	Focusing on horizontal management	Food risk assessment and establishment of standards
Food and Drug Administration	Established at each level of local government at the prefectural level and above	Focusing on horizontal management	Food risk management, general supervision
Ministry of Commerce	Established at each level of local government at the prefectural level and above	Focusing on horizontal management	Safety supervision at the food distribution stage
Administration for Industry and Commerce	Established at each level of local government at the prefectural level and above	Focusing on horizontal management	Safety supervision at the food distribution stage
Administration of Quality Supervision, Inspection and Quarantine	Established at each level of local government at the prefectural level and above	Focusing on horizontal management	Quality supervision at the food production stage
Administration for Enter-exit Inspection and Quarantine	Establishment of provincial (municipality) level governments and local governments of important coastal cities	Central Vertical Management	Supervision of the safety of import and export food

Note 1 Focusing on horizontal management means that each agency is guided by the senior agency to which it belongs and controlled by the local government to which it belongs; central vertical management means that it is primarily guided by the senior agency to which it belongs and is not influenced by the local government

Note 2 The monitoring and management functions of each sector refer to the discussion in the preamble

Source Expanded and created by author

information on food safety is extremely inadequate. It is also persistently pointed out that the level of public confidence in food safety and trust in the system is lower than in Japan.

The issue of low levels of food safety awareness and knowledge among consumers, particularly in rural areas, and room for improvement in the safety management system for agricultural products has become apparent. Information gaps and differences in educational opportunities with urban areas are said to affect consumers' risk perception and ability to respond.

The results of the 2014 survey conducted for this study also highlight the reality regarding consumer risk communication. When respondents were asked if they had ever made a complaint/complaint to a consumer association, approximately 60% of respondents said that they had not. This may reflect the current situation where consumers have difficulty in accessing support organizations even in the event of food safety hazards. Furthermore, when those who answered "yes" were asked how effective they found the complaints/complaints to be after they complained to the consumers' association, approximately 60% gave a negative evaluation. Specifically, many respondents stated that the complaint did not lead to a resolution, that the response was slow, and that no attempt was made to prevent recurrence, indicating that the response and risk communication of the consumer association is not functioning efficiently and effectively.

These results show that the system of consumer risk communication and food safety administration in China is currently not well developed. Although consumer associations themselves play a role in consumer protection, they lack information and education on food safety and consumer participatory institutional design, and lack support for consumers to make appropriate decisions and take appropriate actions. Many experts point to the need to establish efficient and effective risk communication methods, strengthen consumer education and clarify and institutionalize cooperation between government and consumer organizations to fill these gaps.

However, in recent years, in order to respond to trends in domestic and international food safety systems and international standards, the Chinese government has also been working to strengthen relevant institutions and develop standards, and the development of science-based standards and the supervision system is being strengthened, with the National Standard for Food Safety (GB Standard) to be revised on a large scale in 2025. Institutionalization of consumer risk communication is an area for future work, but it is expected that these institutional initiatives will be improved and expanded.

(d) Consumer awareness and perception of food safety

As mentioned above, in recent years, food safety issues in China have led to consumers' anxiety about food safety. The cause of this anxiety can be said to be distrust of the food system. The cause of the problem is due to institutional incompleteness and inadequacy of the spirit of compliance and the law enforcement system. In addition, there are deficiencies in legal norms, administrative guidance, supervision, and management, and a system for promoting consumer awareness of food safety. Scientific food education for consumers has not yet been established, and while awareness of food safety is increasing, knowledge, awareness, and legal awareness of food safety are low. In addition, from the questionnaire survey mentioned in Chapter 3, in response to the question "How is your level of interest in food safety?", 47.5% answered "Very concerned", "Concerned" (31.4%), and "Somewhat concerned" (17.8%). A total of 96.7% of the respondents are concerned about food safety. However, according to the same survey, 60% of respondents answered "No" to the question "Do you understand the safety status of food (information, safety standards, indicators, etc.) when purchasing food?" Furthermore, in response to the question, "How would you respond if there was a food safety hazard happening to you?", more than 60% of the respondents answered "I cried myself to sleep."

5.2.3.2 Recommendations for the Development of Legal and Administrative Systems Related to Food Safety in China Based on Japan's Experience

As mentioned above, due to differences in national circumstances and social backgrounds between Japan and China, it is not possible to directly introduce the experiences of the other parties. However, regarding China's legal and administrative system for food safety, it is believed that policy directions can be presented to lead to solutions to the remaining issues by referring to Japan's experience.

(a) Strengthening the legal system

As shown in Table 5.5, it is necessary to implement detailed legislation that addresses problems in each field with reference to Japan's experience in response to the issues of food safety laws. In addition, it is necessary to revise enacted laws and ordinances in a timely manner in response to changes in the situation related to food safety and related laws and regulations. And, as in Japan, it is necessary to strengthen the relationship, supplementation, and cooperation of laws, laws, and regulations.

(b) Strengthening the administrative system

First, it is necessary to further clarify the responsibilities and authority of each administrative organ and to promote coordination and coordination among each department. In particular, it is necessary to have a mechanism to ensure that the Food Safety Committee, which is the top-level organization, does not become a formality, but functions in practice. Second, with regard to the establishment of local departments, it is desirable to consider and seek reforms to change the management system of the State Food and Drug Inspection Administration, which is a food risk management agency, with reference to the "vertical management" of import/export inspection and quarantine belonging to this division, taking the opportunity of separating the food safety supervision function from the Quality Supervision, Inspection, and Quarantine Division. In other words, it defines the functions of administrative management agencies and grants the authority to legitimate and effective management to the administrative agencies at the end of the administrative body. Third, it is necessary to establish a consumer affairs agency that has a direct relationship with consumers and independent supervisory authority, learning from Japan's experience. Fourth, Japan's experience should be used to ensure greater safety of edible agricultural products, and the monitoring and management systems of relevant laws and administrative agencies should be integrated at the production, processing, distribution, and consumption stages of agricultural products.

(c) Building consumer trust

According to the questionnaire survey mentioned in Chapter 3, the top three influencing the occurrence of food safety problems in China and consumer anxiety and distrust were "decline in morale of food production and management," "inadequacy of food safety administrative system," and "inadequacy of food safety laws." It is believed that the cause of all this is an institutional problem. In other words, the current challenges of food safety administration are consumer neglect and lack of risk communication for consumers. The first is to establish the basic principle of "prioritizing consumers" rather than "prioritizing industry" as in Japan. To this end, it is necessary to develop laws and administrative systems related to food safety, and it is also necessary to establish a consumer affairs agency with administrative authority as a direct response to consumers, and to provide legal guarantees and support for such efforts. Second, it is necessary to emphasize risk communication and collect information, advice, and complaints from consumers, and then communicate effectively and efficiently with consumers. The third is to promote nutrition education in which consumers, business operators and administrative personnel can participate directly, with reference to Japan's nutrition education policy, and to provide moral education to make producers, business operators, administrative departments and citizens aware of compliance, as well as to increase consumers' self-management capacity (knowledge, awareness, etc.) towards food safety.

5.3 China's Food Safety Risk Communication from a Comparison of Japan and China

Through the comparative study discussed above, we presented the institutional issues related to ensuring food safety in China from the perspective of a food safety management system and the direction of the measures. In addition, this section summarizes the challenges of food safety risk communication, which seeks to improve the credibility of food safety in China and the food system through a comparative study between Japan and China.

5.3.1 Challenges in Food Safety Risk Communication

Risk communication, which aims to improve the reliability of food safety in China, is still in its early stages.

(a) From a legal point of view, although the new Food Safety Law (2015) of October 2015 introduced new revisions to the Law to "emphasize risk communication" and "implement social management of food safety and fully fulfill the roles of the media and consumers," it is believed that the effects of these revisions have not yet been sufficiently verified.

(b) From the perspective of the administrative system, the food safety administration has traditionally neglected consumers (protection) rather than producers, so consumer administration centered on consumers has not yet been established. There is no national administrative structure that supports centralization to promote food safety risk communication for consumers, and there is no administrative structure that clearly divides the responsibilities of each ministry and agency (see Figs. 5.5 and 5.6).
(c) From the practical aspect of food safety risk communication, as a way to promote risk communication that bears responsibility, such as the collection, analysis, and presentation of information related to food safety, supervision and management of food labeling, safety guidance for food consumption, and food safety education, "information processing on cases (Consumer Policy Division, Safety Division) and public safety education (Consumer Affairs Information Division)" as in Japan. It has not yet been clearly established that "through the National Consumer Affairs Center, which is under the jurisdiction of the Regional Cooperation Division, we will collect information, advice, and complaints from consumers, provide information and present safety to consumers, and work to ensure the safety and trust of food products." There are insufficient efforts to educate the public to ensure that consumers have basic knowledge about food safety and to raise awareness of food risks. The development of a food labeling system and the dissemination of a food traceability system are also insufficient, and disclosure of information on food safety is required. Since consumers do not have the ability to judge the situation or respond to false or misleading reports about food safety, it is necessary to build a scientific and effective food safety information sharing system and develop methods for collecting and disseminating food safety information to consumers.

5.3.2 Summary: The Top Priority of Food Safety Policy is the Protection of People's Health and Lives

As shown in Table 5.10, the food safety legal systems of China and Japan both share the same priority of protecting the health of the public, but there are clear differences in their institutional design. In China, a system of integrated regulation from production to distribution and consumption has been established, centered on the Food Safety Law enacted in 2009, which is a "regulatory and enforcement-oriented" system characterized by a strict supervision system and high penalties. This is due to the growing need for the State to take the lead in strongly ensuring safety against the backdrop of serious food accidents, such as the Sanlu milk powder incident.

In Japan, on the other hand, the Food Safety Basic Act enacted in 2003 is positioned as a principle law and adopts a "risk analysis-oriented" system with a clear division of labor between risk assessment, risk management and risk communication. It is characterized by the establishment of the Food Safety Commission as an

Table 5.10 Comparison of food safety legislation between China and Japan

Point of view	China	Japan
Basic legislation	Food Safety Law (enacted 2009, amended several times)	Food Safety Basic Act (enacted in 2003)
The nature of the basic legislation	Substantive legislation including regulation, supervision and penalties	Basic law setting out principles and framework
Background	A number of serious food accidents, including the 2008 Sanlu milk powder incident	Decline in trust in food safety administration due to BSE and other problems
Basic principles	Protection of people's lives and health, strict state supervision	Priority on public health protection, emphasis on scientific evidence
Risk analysis	Integrated as a system by the Food Safety Law	Clear separation of risk assessment, management and communication
Risk assessment bodies	Relevant departments of the State Council (e.g. National Health Commission)	Food Safety Commission (Cabinet Office, highly independent)
Production stage	Agricultural Product Quality Safety Law, Agricultural Law, etc.	Food Sanitation Act, JAS Act, Fertilizer Control Act, etc.
Processing/manufacturing stage	Food Safety Law, Product Quality Law	Food Sanitation Act (sanitation management in line with HACCP is mandatory)
Distribution and sales stage	Distribution Food Safety Supervision and Management Act, etc.	Food Sanitation Act, Act against Unjustifiable Premiums and Misleading Representations
Import/export control	Law on Import and Export Commodity Inspection, Animal and Plant Quarantine Law	Food Sanitation Act, Plant Protection Act, Animal Quarantine Service
Food labelling	National Food Safety Standards (GB Standards), Product Quality Law	Food Labelling Act (under the jurisdiction of the Consumer Affairs Agency)
Consumer protection	Law on Protection of Consumer Rights and Interests	Basic Act on Consumer Policies, Consumer Safety Act
Penalty features	Severe penalties, including high fines and business suspensions	Focus on administrative guidance and improvement orders, criminal penalties for malicious cases

(continued)

Table 5.10 (continued)

Point of view	China	Japan
Risk communication	Government-led, intensified in recent years	Emphasis on interactivity between the state, businesses and consumers
Relationship with nutritional education	Food safety education is part of the policy	Systematically promoted by the Basic Act on Shokuiku (Food and Nutrition Education)

Source Compiled by the author from publicly available materials

independent risk assessment body and an emphasis on policy decisions based on scientific evidence. In addition, through the Food Labelling Act and the Basic Act on Food Education, a system is in place to promote consumer understanding and independent decision-making.

Thus, while China emphasizes food safety assurance based on strong state regulation and penalties, Japan has established a system based on scientific evaluation and trust-building between relevant actors, and the food safety legal systems of the two countries have contrasting features reflecting differences in social background and policy philosophy.

Both China and Japan regard the protection of the health and lives of their citizens as the top priority of their food safety policies, but there are significant differences in the design and operation of their systems. China has chosen to ensure food safety by "strengthening management and regulation", with a strong state supervision and penalty system in place since the enactment of the Food Safety Law in 2009, triggered by serious food accidents such as the Sanlu milk powder incident. This has clarified legal responsibilities for food businesses and enforced compliance with safety standards by imposing severe penalties, including hefty fines and suspensions of operations.

In contrast, Japan has a system design that delegates science-based risk assessment to an independent body, separates and clarifies risk management and risk communication, and emphasizes transparency and collaboration between the actors involved. Such a "risk analysis-oriented" approach is effective in terms of building consumer confidence in food safety and encouraging flexible and appropriate responses. A further distinctive feature is the promotion of consumer education through the Basic Law on Food Education, which strengthens the social basis for food safety.

The first suggestion that can be drawn from these contrasting approaches is that food safety policy should not just be about strengthening regulations, but that a comprehensive combination of science-based risk assessment and effective communication with consumers is essential for ensuring sustainable food safety. Second, increasing the transparency and independence of the system is crucial for gaining consumer confidence, and this should be taken into account in the design of the legal system. Third, positioning nutrition education and consumer education as an important pillar of policy is expected to increase consumer literacy and contribute to the development of a common social understanding of food safety.

Therefore, in order to effectively ensure food safety, in addition to strengthening legal regulations, it is necessary to promote multifaceted policies in an integrated manner, such as the development of a scientific risk assessment system, improving administrative transparency and enhancing consumer education, and the experiences of China and Japan provide important insights that can contribute to their design.

5.4 Policy Implications: Rebuilding Institutional Trust and Food Safety Governance from a Japan–China Comparative Perspective

The policy implications derived from the comparative analysis of Japan and China in this chapter are directly related, in particular, to the reconstruction of "institutional trust (governance level)" within the three-layer trust model proposed in this book. This book conceptualizes trust in food safety as an integrated model consisting of "psychological trust (individual level)," "institutional trust (governance level)," and "structural and sustainability-based trust (societal level)." The five policy implications discussed below, through a comparison with Japan's institutional experience, not only indicate the future direction of food safety governance in China but also clarify, from the perspective of institutional design, the mechanisms through which institutional trust is formed.

First, there is a need to shift from regulation-centered governance to trust-building governance. In China, the Food Safety Law has been progressively improved and penalties have been strengthened. However, as the Japanese case demonstrates, the strictness of legal institutions alone is insufficient to secure consumer trust. In Japan, repeated food safety incidents prompted not only regulatory reinforcement but also a transition toward institutional designs emphasizing information disclosure and accountability, thereby gradually accumulating public trust. This comparison suggests that, in China, strengthening regulations must be accompanied by a transition toward "trust-building governance" that emphasizes transparency and participation.

Second, the institutionalization and advancement of risk communication are essential. In Japan, a tripartite framework integrating "risk assessment, risk management, and risk communication" has been established, centered on the Food Safety Commission, ensuring systematic information sharing and dialogue with consumers. In contrast, China still faces challenges such as insufficient information provision and one-way communication. Therefore, China needs to enhance consumer-oriented information dissemination, appropriately utilize social media and digital platforms, and improve credibility through the involvement of third-party institutions, drawing on Japan's institutional experience to advance its risk communication practices.

Third, it is necessary to integrate administrative systems and clarify responsibilities. In Japan, functional divisions—risk assessment (Food Safety Commission), risk management (relevant ministries), and risk communication (Consumer Affairs Agency)—are relatively clearly defined, forming a coherent governance system through inter-agency coordination. In contrast, China continues to face issues such as overlapping authority among departments and ambiguous accountability. Thus, it is essential to establish a unified supervisory system covering the entire "farm-to-table" process, along with a clear delineation of responsibilities among relevant actors.

Fourth, strengthening traceability and information transparency is critical. In Japan, traceability systems for products such as beef and rice, along with well-developed food labeling systems, have enabled "visible safety" and formed the foundation of consumer trust. In China, however, institutionalization of information disclosure and traceability remains incomplete in some areas. This comparison indicates that China should strengthen the institutional mechanisms for recording, tracking, and disclosing information across all stages of production and distribution, thereby building a system that accumulates trust in normal times. This implies a shift from reactive governance to preventive and transparency-oriented governance.

Fifth, enhancing consumer capability and promoting food education are indispensable. In Japan, under the Basic Law on Shokuiku (food education), consumer education has been promoted to improve knowledge and understanding of food safety. In contrast, China still faces disparities in consumers' ability to understand information and perceive risks. Therefore, China should enhance consumers' self-management capacity through the dissemination of scientific knowledge and risk education, while also promoting rational consumption behavior.

In summary, the most important implication derived from the China–Japan comparison is the need to reconceptualize food safety policy from a "problem of regulation" to a "problem of the structure of trust." As illustrated by Japan's experience, constructing a "trust-building food safety governance" system that integrates the interactions among institutions, administration, information, and consumers constitutes the core challenge for achieving a sustainable food safety system in China.

Acknowledgements This chapter is an extensively revised, expanded, and updated English version of the author's article, "An Institutional Study on Ensuring Food Safety in Japan: Focusing on Legal Frameworks and Administrative Systems," published in *Policy Science*, Vol. 22, No. 1, The Association of Policy Science, Ritsumeikan University, 2014.

References

Bian, Y. (2012). An overview of Chinese law on food safety. *Frontiers of Law in China*, 7(1), 91–112. https://doi.org/10.3868/s050-001-012-0005-0

Food and Agriculture Organization of the United Nations. (2024). *Food supply (kcal per capita per day): Japan*. FAOSTAT. https://www.fao.org/faostat/en/#data/FBS

Food Safety Commission of Japan. (2003). Food Safety Basic Act (Act No. 48 of May 23, 2003). Government of Japan. https://www.fsc.go.jp/english/basic_act/fs_basic_act.pdf

Hu, J. (2007). *Hold high the great banner of socialism with Chinese characteristics and strive for new victories in building a moderately prosperous society in all respects: Report to the 17th National Congress of the Communist Party of China*. Beijing: People's Publishing House.

Japan Finance Corporation. (2014, September 11). Heisei 26-nendo kamihanki shōhisha dōkō chōsa [Consumer trend survey: First half of fiscal year 2014]. https://www.jfc.go.jp/

Japan Finance Corporation. (2024, September 12). Shōhisha dōkō chōsa (Reiwa 6-nen 7-gatsu) [Consumer trend survey, July 2024]. https://www.jfc.go.jp/n/findings/pdf/topics_240912a.pdf

Japan Food Service Association. (2023). Gaishoku sangyō shijō dōkō chōsa [Survey of market trends in the food service industry]. https://www.jfnet.or.jp/data/industry.html

Japan Tourism Agency. (2025). White paper on tourism in Japan, 2024: Summary. Ministry of Land, Infrastructure, Transport and Tourism. https://www.mlit.go.jp/kankocho/content/001767069.pdf

Li, D., & Minamiishi, A. (2013). Chūgoku ni okeru shokuhin anzen gyōsei no shinkyokumen oyobi sono kadai: Kokumuintō kikō kaikaku kinō tenkan-an to Nihon no keiken o chūshin to shite [New developments and challenges in China's food safety administration: State Council institutional reform and lessons from Japan]. Kyushu University Journal of Agricultural Science, 68(2), 35–47.

Liu, P., & Zhang, S. (2015). Toward regulatory federalism: A study on the vertical power distribution of China's food safety regulatory system. *Journal of Huazhong Normal University(Humanities and Social Sciences), (01)*, 28–34.

Ministry of Agriculture, Forestry and Fisheries. (2013). Shokuryō, nōgyō, nōson hakusho 2012 [Food, agriculture and rural areas white paper 2012]. Government of Japan. https://www.maff.go.jp/

Ministry of Agriculture, Forestry and Fisheries. (2021). Food, agriculture and rural areas white paper 2021. Government of Japan. https://www.maff.go.jp/j/wpaper/w_maff/2021/pdf/01overview.pdf

Ministry of Health, Labour and Welfare. (2023). Shokuhin no anzen kakuho ni muketa torikumi [Measures to ensure food safety]. Government of Japan. https://www.mhlw.go.jp/

National People's Congress of the People's Republic of China. (2015). Food Safety Law of the People's Republic of China (2015 revision). http://www.npc.gov.cn/

OECD. (2024). Gross domestic product (GDP). OECD Data. https://data.oecd.org/gdp/gross-domestic-product-gdp.htm

State Council of the People's Republic of China. (2019). *Opinions on deepening reform and strengthening food safety work*. Beijing: State Council.

Wang, F., Zhou, W., & Nakagami, K. (2014). Nihon ni okeru shokuhin anzen kakuho shisutemu no seido bunseki [Institutional analysis of food safety assurance systems in Japan]. Policy Science, 22(1), 23–37.

Wang, F., Zhou, W., Qian, X. & Nakagami, K. (2016). Risk communication for improving consumer trust in food safety in China: An empirical analysis of food safety information. *Social System Studies, 33*, 55–85.

World Bank. (2026). *GDP per capita (current US$)* [World Development Indicators dataset]. The World Bank. https://data.worldbank.org/indicator/NY.GDP.PCAP.CD

Wu, Y.-N., Liu, P., & Chen, J.-S. (2018). Food safety risk assessment in China: Past, present and future. *Food Control*, 90, 212–221. https://doi.org/10.1016/j.foodcont.2018.02.049

Chapter 6
Structural Transformation of Food Safety Governance: The Formation of Institutional Trust and Institutional Public Goods

Weisheng Zhou, Fengyang Wang, and Xuepeng Qian

6.1 Problem Awareness and Analytical Framework: From "Regulation" to "Governance", What Does the Institutionalization of Trust Mean?

The central concern of this chapter is to reposition trust in food safety within an institutional and governance-oriented context, rather than reducing it to a matter of consumer psychology. In the previous chapter, Chap. 5, the challenges and directions for improvement in China's food safety system were examined through a Japan–China comparison, and the importance of reconstructing institutional trust at the governance level was particularly demonstrated. Building on that discussion, this chapter seeks to illuminate, in a more theoretical and structural manner, the mechanisms through which institutional trust is formed.

The empirical analyses conducted in the preceding chapters confirmed that consumer trust in China is mediated by multiple factors, including risk perception, trust in information, and evaluation of institutions. However, these factors are not confined to individuals' subjective attitudes; they are equally shaped by the design of social institutions and the consistency of administrative enforcement. In other words, trust is not merely a psychological phenomenon but a product of institutional structures.

W. Zhou (✉)
College of Policy Science, Ritsumeikan University, Kyoto, Japan
e-mail: zhou@sps.ritsumei.ac.jp

F. Wang
Xi'an Peihua University, Xi'an, China

X. Qian
Graduate School of Global Environmental Studies, Sophia University, Tokyo, Japan
e-mail: qianxp@sophia.ac.jp

W. Zhou (ed.), *Food Safety in China*, https://doi.org/10.1007/978-981-92-3741-8_6

Theoretical studies on trust have understood trust as a structure of expectations under uncertainty. Luhmann (1979) positioned trust as a mechanism for reducing social complexity and argued that institutional stability enhances the predictability of expectations. Giddens (1990) also pointed out that trust in abstract systems in modern society depends on expert knowledge and institutional legitimacy. Furthermore, Beck's (1992) theory of risk society reveals that, as modernization advances, institutions themselves can become sources of risk. These theoretical contributions provide a lens for understanding food safety not merely as a matter of technical management, but as a problem of institutional trust.

The importance of institutional trust has also been internationally recognized in the field of food safety. The World Health Organization (2022) points out that the core of ensuring food safety lies in scientific risk assessment and transparent risk communication. The Food and Agriculture Organization of the United Nations (2023) also reports that the strengthening of governance capacity is a decisive factor for the sustainability of food safety. In other words, unless institutional transparency, accountability, and consistency in enforcement are ensured, consumers' subjective sense of security cannot be stably formed.

In China, a succession of serious food safety incidents since the 2000s has rendered the instability of institutional trust increasingly visible. One turning point was the enactment of the Food Safety Law in 2009, through which a shift from a fragmented supervisory system to an integrated management system was pursued. Furthermore, the 2015 amended law introduced strengthened penalties, clarified corporate responsibility, and established the principle of whole-process management, thereby advancing the development of an institutional framework (Standing Committee of the National People's Congress, 2015). These reforms should be understood not merely as legislative measures aimed at accident prevention, but as a broader political and social project to reconstruct institutional trust.

However, institutional reform does not automatically translate into the restoration of trust. Trust changes dynamically not only through institutional design, but also through the enforcement process, the quality of information disclosure, and interactions between administration and society. In particular, the advancement of digitalization has dramatically expanded the speed and scope of information transmission, while also accelerating the spread of uncertainty and rumors. For this reason, food safety governance must be understood as a complex structure encompassing legal institution-building, administrative capacity development, corporate behavioral discipline, and the institutionalization of citizen participation.

In this chapter, food safety reliability is understood as a social trust resource characterized by non-excludability and non-rivalry, sustained by institutional frameworks such as government regulatory systems, corporate quality management, information disclosure, and social monitoring. Although formed through the accumulation of efforts by individual actors, it is widely shared across society and functions as an "institutional public good" that supports the stability and sustainability of the food market. Therefore, food safety reliability is not limited to subjective perception or market evaluation; rather, it possesses a public character generated and maintained through the interaction of institutions, norms, and technology. For this reason, misconduct or information concealment by some actors entails negative externalities that damage overall trust, and public intervention and collaborative governance are indispensable for its provision and maintenance.

Trust is not the mere accumulation of evaluations of individual companies or administrative officials, but rather a reflection of expectations regarding the institutional stability of society as a whole. The more predictable institutions become, the more easily consumers can rationally accept risk. Conversely, when institutional enforcement is arbitrary or information disclosure is opaque, psychological anxiety is amplified.

Based on the above, the problem awareness of this chapter can be summarized in the following question: Through what mechanisms can the structural transformation of food safety governance in China form institutional trust? To answer this question, this chapter systematically examines the development of legal and institutional reforms, the reorganization of administrative organizations, the strengthening of corporate responsibility, and the introduction of digital technologies, and analyzes their structural effects on trust formation.

6.2 Development of China's Food Safety Legal System and Trends in Institutional Reform

Food safety governance in China has undergone repeated institutional transformations since the 2000s, triggered by serious incidents. This section traces these structural changes, beginning with the enactment of the Food Safety Law in 2009, the regulatory strengthening brought by the 2015 amended law, and the administrative restructuring after 2018. These reforms are not mere legal-technical revisions; they signify a shift from fragmented to integrated risk management, and further from an "ex post punishment-oriented" model to a "whole-process management-oriented" model.

6.2.1 Enactment of the 2009 Food Safety Law: From Fragmented Management to an Integrated Framework

In the early 2000s, China's food safety supervision system was dispersed among multiple departments, including agriculture, health, industry and commerce, and quality supervision, leaving lines of responsibility unclear. Following the 2008 melamine contamination incident and the resulting erosion of public trust, China enacted the Food Safety Law in 2009. The law established a more comprehensive legal framework covering food production, distribution, risk assessment, safety standards, and regulatory supervision (Standing Committee of the National People's Congress, 2009).

The law was built around several pillars, including the national unification of food safety standards, the introduction of a risk assessment system, the strengthening of food additive management, and the clarification of enterprises' primary responsibility. In particular, the establishment of a risk assessment committee represented an attempt to institutionalize regulation based on scientific evidence and indicated a direction toward greater consistency with international frameworks. This trend is also consistent with the "evidence-based food safety system" advocated by the World Health Organization (World Health Organization, 2022).

Nevertheless, although the 2009 law marked the starting point for institutional integration, it left unresolved issues such as disparities in local enforcement capacity and overlaps in supervisory authority. The influence of a fragmented administrative culture remained strong, and a gap persisted between institutional design and implementation.

6.2.2 The 2015 Amended Food Safety Law: Strengthened Penalties and the Principle of Whole-Process Management

The 2015 amended Food Safety Law has been widely characterized as the "strictest in history." The core of the amendment consisted of a substantial strengthening of penalties, the clarification of corporate responsibility, and the explicit codification of the principle of "whole-process management" from farm to table (Standing Committee of the National People's Congress, 2015).

First, fines imposed on violating enterprises were changed to a sales-based method, thereby greatly strengthening economic deterrence. In addition, enterprises were required to establish internal quality management systems and introduce traceability systems, shifting the policy emphasis from ex post punishment to ex ante prevention. Furthermore, the amendment incorporated whistleblower protection and strengthened social supervision, showing the emergence of citizen-participatory governance.

This amendment should be read not merely as regulatory tightening, but as an institutional signal aimed at restoring public trust. From the perspective of trust theory, strengthened penalties serve the function of "maintaining order through deterrence," while the institutionalization of traceability and information disclosure supports institutional trust by "improving predictability" (Giddens, 1990). However, penalty-dependent institutions have limits in forming long-term trust, and ensuring fairness and transparency in enforcement is indispensable.

6.2.3 Administrative Restructuring After 2018: Establishment of the State Administration for Market Regulation

Through the state institutional reform of 2018, food safety supervision functions were integrated into the newly established State Administration for Market Regulation (SAMR). As a result, functions that had previously been dispersed among industry and commerce, quality supervision, and food and drug supervision were unified, thereby aiming to eliminate overlaps in supervisory authority and improve administrative efficiency (State Council of the People's Republic of China, 2019).

This restructuring was intended to overcome vertical administrative segmentation and strengthen risk management capacity, and it is characterized by the reinforcement of centralized control. At the same time, the clarification of supervisory responsibilities for local governments was also advanced. The readjustment of central–local relations has important significance within China's distinctive governance structure.

Moreover, in recent years, the introduction of digital supervisory systems has progressed, and systems for risk prediction through the monitoring of online transactions and big data analysis are being constructed. This is also consistent with the direction emphasized by the Food and Agriculture Organization of the United Nations regarding "capacity building and institutional infrastructure development for food control systems" (Food and Agriculture Organization of the United Nations, 2023).

6.2.4 Evaluation of Institutional Reform and Remaining Challenges

Based on the above developments, China's food safety legal system can be understood as having undergone three stages of reform: first, the integration of the legal system; second, the strengthening of penalties and the introduction of preventive principles; and third, the reorganization of the administrative system. These represent important progress toward restoring institutional trust.

However, challenges remain. First, there is an imbalance in enforcement capacity at the local level. Second, there is the issue of increasing compliance costs for small and medium-sized enterprises. Third, citizens remain skeptical about the substantive transparency of information disclosure. Trust depends not on the mere existence of institutions, but on their effectiveness and fairness (Beck, 1992).

Therefore, future institutional reform requires not merely the addition of laws and regulations, but a comprehensive approach that includes the equalization of enforcement capacity, the strengthening of third-party audits, and the institutionalization of social participation. The institutional developments organized in this section form the premise for the analysis of the transformation of governance structures and mechanisms of trust formation examined in the next section.

6.3 The Tripartite Relationship Among Government, Enterprises, and Consumers in Food Safety Issues

6.3.1 From an Administrative Regulation-Oriented Model to a Collaborative Governance Model

Food safety governance in China has long operated through a government-led, administrative regulation-oriented model. This is a form of "administrative regulation-oriented" governance in which the government formulates laws, regulations, and standards, and unilaterally requires enterprises to comply with them. However, as food safety risks have diversified and supply chains have grown increasingly complex, it has become impossible for the government alone to oversee the entire system. Collaboration with diverse stakeholders—enterprises, consumers, and civil society—has therefore become indispensable. Against this backdrop, the transition toward a "collaborative governance model" has received growing emphasis in both policy documents and academic research.

There is a growing consensus that food safety is not merely an object of administrative control, but that sustainable safety can be achieved only when the state, enterprises, society, and citizens each fulfill their respective roles. Wu et al. (2022) point out that, in the social co-governance of food safety in China, food producers, distributors, and retailers, as well as the government, social organizations, news media, and citizens, need to share responsibilities. In other words, media organizations and ordinary citizens responsible for information dissemination are also positioned as important actors in risk management.

This concept of collaborative governance is clearly reflected in the 2015 amendment to the Food Safety Law. The law positioned the establishment of traceability systems and quality management systems by enterprises as responsibilities of business actors, thereby clarifying the division of responsibility between administration and market actors (Standing Committee of the National People's Congress, 2015). Furthermore, by institutionalizing the strengthening of social supervision and information disclosure, the law expanded the scope for citizens and enterprises to become involved in the supervisory process. As a result, the shift from a government-led regulatory system to a pluralistic governance framework that includes voluntary corporate compliance and social participation has progressed.

At the institutional level, policy coordination bodies such as food safety committees have been established, and cross-ministerial consultations among relevant ministries and agencies, as well as the formulation of unified guidelines, have been promoted. Such integrated coordinating functions play a role in managing complex risks that are difficult for a single department to address, while institutionally promoting information sharing and consultation with enterprises and social actors.

This governance transformation has also had a major impact on corporate behavior. Many food enterprises have obtained international quality standards and third-party certifications, such as ISO 22000 and FSSC 22000, and have sought to build market trust through transparent quality management. These voluntary initiatives function in a complementary manner with government regulation and contribute to the reconstruction of consumer trust. Certification systems are important mechanisms through which enterprises demonstrate externally not only legal compliance, but also the transparency and consistency of their quality management systems.

The role of consumers is also undergoing significant change. With the spread of the internet and social networking services, the sharing of risk information has become generalized, and social pressure functioning as a supervisory mechanism over governments and enterprises has increased. In addition, the weight of responsibility in food safety tends to be distributed in the order of food producers, distributors, and retailers; the government; social organizations; news media; and citizens (Wu et al., 2022). This finding indicates the need for consumers to participate actively in the risk management process and also confirms that collaborative governance presupposes the roles of diverse actors.

At the policy level, the government has developed institutional channels for receiving consumer complaints and reports, while also promoting scientific understanding of risk through education and information provision. Through these measures, consumers are increasingly being positioned not as "passive recipients," but as actors who actively participate in the formation of food safety.

As described above, food safety governance in China has gradually shifted from the traditional model centered on administrative regulation toward collaborative governance that includes the strengthening of corporate responsibility, the participation of social organizations, and consumer involvement. This transition does not merely represent an expansion of institutions; rather, it indicates a direction toward improving institutional trust through the design of responsibility sharing and interaction among stakeholders.

6.3.2 Corporate Compliance and CSR: Enterprises as Actors in Trust Formation

Within the collaborative governance model, enterprises are positioned not merely as objects of regulation, but as active agents of trust formation. In particular, the 2015 amended Food Safety Law clearly positioned enterprises as the "primary responsible actors" and stipulated the establishment of internal management systems, the construction of traceability systems, and obligations for self-inspection (Standing Committee of the National People's Congress, 2015). This marks a shift away from administration-dependent management and toward a stronger emphasis on internal corporate governance.

First, the institutionalization of compliance systems has progressed. Many food enterprises have introduced international standards such as HACCP, Hazard Analysis and Critical Control Point, which is a preventive hygiene management method for ensuring food safety, and ISO 22000, thereby systematizing risk management. The Codex Alimentarius Commission (2022), in its General Principles of Food Hygiene, positions HACCP-based approaches as important methods for improving food safety. In China, mainly among large enterprises, the acquisition of third-party certification has advanced, and quality assurance has come to be positioned as part of brand strategy. This reflects a strategic reorientation in which trust is regarded not as a compliance cost, but as a source of competitive advantage.

Second, corporate social responsibility, or CSR, has become increasingly important. As Giddens (1990) points out, trust in modern society is based on dependence on abstract systems, and contemporary enterprises are actors that constitute the core of such systems. In China as well, food enterprises are increasingly publishing social responsibility reports and disclosing transparency from raw-material procurement to distribution. CSR provides enterprises with ethical legitimacy that extends beyond legal compliance and contributes to the formation of institutional trust.

Third, challenges remain for small and medium-sized enterprises. The introduction of advanced quality management systems creates a financial burden, and the sophistication of institutions may promote market concentration. On this point, capacity building through government technical support and subsidy systems is indispensable (World Health Organization, 2022). Collaborative governance is not merely a matter of sharing responsibilities; it must also be accompanied by mechanisms for correcting disparities in capacity among actors.

6.3.3 Responsibilities of Platform Enterprises: Food Safety Governance in the Digital Age

The rapid expansion of electronic commerce has greatly transformed the structure of food distribution. While the spread of online sales has improved convenience, it has also generated new risks, such as the opacity of seller information and the circulation of counterfeit or misrepresented products.

First, platform operators' legal responsibilities have been progressively clarified. The 2015 revised Food Safety Law and the subsequent 2016 regulatory measures require online food-trading platforms to verify and register sellers, review relevant business licences, prevent and report illegal conduct, suspend platform services in cases of serious violations, and retain transaction-related information (Standing Committee of National People's Congress, 2015; China Food and Drug Administration, 2016). As a result, platform enterprises have come to be positioned not merely as intermediaries, but as jointly responsible actors. In addition, the State Administration for Market Regulation, or SAMR, has strengthened supervision of online transactions and institutionalized the primary responsibilities of online business operators, information preservation, consumer protection, and the maintenance of transaction order (State Administration for Market Regulation, 2021).

Second, the use of big data and risk prediction has advanced. Based on enormous volumes of transaction data, platform enterprises are able to detect fraudulent transactions and risks at an early stage. In its report on future issues in food safety, the Food and Agriculture Organization of the United Nations emphasizes that technological innovation and digitalization create new possibilities for food safety management, while also stressing the importance of early risk identification and institutional responses (Food and Agriculture Organization of the United Nations, 2022). At the same time, the need for institutional design that ensures algorithmic transparency and data authenticity has also been pointed out.

Third, consumer participation through evaluation systems plays an important role. Online reviews and rating systems promote information sharing among consumers and function as "social supervision"; however, problems such as false reviews and information manipulation also exist. For this reason, platforms are required to achieve both supervision and self-discipline.

As described above, the tripartite relationship among government, enterprises, and consumers is transforming from a traditional vertical relationship into an interdependent network. Institutional trust is strengthened when administrative regulation provides the foundation, while corporate compliance and CSR, consumer participation, and platform responsibility function in a complementary manner. This collaborative governance model repositions trust not as a "by-product of regulation," but as an "objective of institutional design," and symbolizes the structural transformation of China's food safety system.

6.3.4 The Role of Consumers and the Feedback Function

As clarified in Chaps. 3 and 4, food safety issues in China should be understood not merely as matters of institutional design or corporate behavior, but as "issues of trust" that are strongly shaped by consumers' risk perception and information environment. In other words, consumers are not passive recipients of risk, but important constituent actors in food safety governance. Their judgments of trust and their actions perform a feedback function that shapes the responses of both government and enterprises.

First, as the analysis in Chap. 3 shows, consumer trust in food safety depends not only on objective levels of safety, but also heavily on subjective risk perception and the reliability of information. In China in particular, due to memories of past food safety incidents and the influence of mass media reporting, there exists a "gap between perception and reality," in which anxiety persists even though safety has objectively improved. This phenomenon is consistent with risk society theory, which points to the increase of uncertainty in modern society and the socially constructed nature of risk perception (Beck, 1992). It also indicates that trust formation depends not simply on the amount of information, but on its quality and reliability.

Furthermore, the empirical analysis in Chap. 4 revealed that the reliability of food safety information plays a decisive role in the formation of consumer trust. While trust in government agencies and experts is relatively high, trust in information issued by enterprises is low, and external information and mass media may amplify distrust. This result is also consistent with Akerlof's (1970) framework, which discusses the problem of market trust arising from information asymmetry, namely a condition in which, in transactions or interactions, one actor possesses more information, or information of higher quality, than another actor, and this informational gap affects decision-making and outcomes. In addition, although the spread of social media has accelerated information circulation, it does not necessarily contribute to the improvement of reliability; rather, it also has the aspect of structurally reinforcing distrust through the spread of misinformation (van Dijck, 2013).

Based on the above empirical findings, consumers perform three roles in food safety governance. First, as the "ultimate evaluators of trust," consumers judge the legitimacy of the actions of government and enterprises. Second, purchasing behavior and risk-avoidance behavior, such as shifting toward imported foods, influence corporate behavior through market mechanisms. Third, information dissemination through SNS, word of mouth, and reporting systems functions as a feedback mechanism that encourages the strengthening of government regulation and the improvement of corporate compliance.

The tripartite relationship model among government, enterprises, and consumers presented in this section understands such interactions as a "feedback loop" (Fig. 6.1). The government forms the foundation of institutional trust through institutional design and information disclosure, while enterprises participate in trust formation in the market through quality management and information dissemination. Meanwhile, consumers make trust judgments based on risk perception and provide feedback to the government and enterprises through the actions that result from those judgments. In this cyclical structure, trust is understood not as a fixed state, but as a process dynamically generated and reorganized through social interaction (Luhmann, 1979).

Furthermore, this tripartite relationship closely corresponds to the three-layered trust model—psychological, institutional, and structural—presented in this book (Table 6.1). Specifically, consumers' risk perception and trust judgments are located at the level of psychological trust, or the micro level; government institutional design and regulation constitute institutional trust, or the meso level; and social structures, including platforms and the information environment, form structural trust, or the macro level. What is important is that consumers, while located at the micro level, function as "mediating actors" who influence institutions and structures through their actions. This multilayered feedback structure has been further strengthened by the advancement of digitalization.

This feedback structure is also consistent with the process through which trust functions as a mechanism for reducing social complexity (Luhmann, 1979). In other words, when trust is established, the stability of markets and institutions is maintained. Conversely, when distrust expands, it functions as pressure that prompts institutional reform and the reorganization of corporate behavior.

Therefore, in order to enhance food safety governance in China, it is indispensable to position consumers not merely as objects of protection, but as actors in "co-governance." This transition signifies a shift from a one-directional regulation model led by the government to a trust formation system based on interaction among the

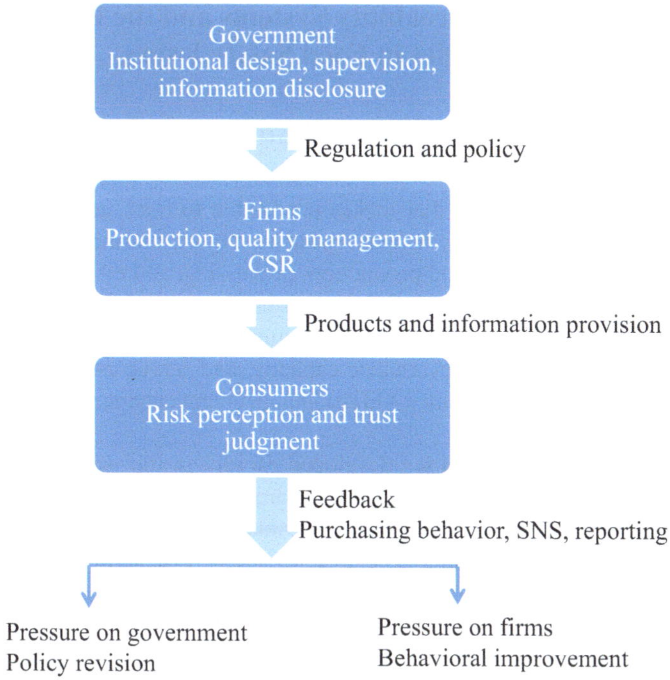

Fig. 6.1 Feedback loop model among government, firms, and consumers (conceptual diagram, created by the author)

Table 6.1 Correspondence between the relationship among government, firms, and consumers and the three-layered trust model

Layer of trust	Actor	Content
Psychological trust (micro)	Consumers	Risk perception, anxiety, subjective trust
Institutional trust (meso)	Government and institutions	Regulation, supervision, information disclosure
Structural trust (macro)	Market and social structure	Platforms, information environment
Integrative mechanism	Tripartite relationship	Feedback, co-governance

three actors. To achieve this, it is necessary to institutionalize transparent information disclosure and two-way risk communication by the government, reliable information dissemination by enterprises, and the improvement of consumers' information literacy and the promotion of their active participation (Frewer, 2004). As described above, the tripartite relationship among government, enterprises, and consumers should be reformulated not merely as a division of roles, but as a dynamic governance mechanism that mediates the three-layered trust relationship of psychology, institutions, and structure.

6.4 Development of Traceability Systems and the Role of Digital Technologies in Food Safety Issues

6.4.1 Development of Traceability Systems

Traceability refers to a system that makes it possible to track and verify product history information concerning where and what kind of processing has been carried out throughout each stage of food production, processing, and distribution (Tomy's Corp., 2020). In responding to food safety issues in China, the development of traceability systems represents a core policy instrument linking digitalization with the strengthening of institutional governance capacity, and serves as an indispensable foundation for food safety assurance and risk management enhancement (Tang et al., 2015).

In China, institutional responses developed rapidly in the wake of a series of food safety incidents, including the 2008 melamine contamination incident. The culmination of this development was the 2015 amended Food Safety Law, which clearly stipulated the establishment of a "food safety traceability system" at the national level. Specifically, it obligated food production and distribution operators to ensure traceability through the use of informatized means, and institutionalized the development of an integrated traceability system through cooperation among government departments (Standing Committee of the National People's Congress, 2015).

This institutional framework signifies a shift from the traditional model of ex post supervision and enforcement-oriented regulation to risk-preventive and process-management-oriented governance. In addition to traceability, the amended Food Safety Law comprehensively introduced risk assessment, a recall system, and strengthened information disclosure, with the aim of comprehensively enhancing the institutional capacity of food safety governance (Standing Committee of the National People's Congress, 2015; Liu & McGuire, 2019).

Furthermore, China's traceability system is characterized by strong government-led institutional design and the introduction of digital technologies. Specifically, mechanisms that make use of barcodes, QR codes, electronic certification systems, and other tools have been introduced in various regions to manage information such as places of food production, processing histories, and distribution routes in an integrated manner. The development of such information infrastructure makes it possible to ensure food quality and rapidly trace causes when accidents occur, thereby contributing to the restoration of consumer trust (Duan et al., 2017).

In recent years, it has also become clear that combinations of policy instruments, or policy mixes, promote the introduction of traceability. For example, empirical studies have shown that combining strengthened sampling inspections with the institutionalization of information disclosure increases business operators' incentives to adopt traceability systems (Zhou et al., 2022).

Nevertheless, despite institutional development, several challenges have been pointed out at the implementation stage. First, China's food supply system includes many small-scale producers, making the costs of information collection and standardization high. Second, disparities exist among local governments in enforcement capacity and technological infrastructure, making it difficult to construct a nationally uniform traceability system. Third, even if institutions have been developed, problems remain in ensuring the reliability and interoperability of actual data (Duan et al., 2017).

Based on the above, the development of traceability systems in China must be understood not merely as the introduction of technology, but as a process of upgrading governance capacity through the integration of "institutions × technology × market incentives." In other words, digitalized traceability transforms food safety from "responses to individual incidents" into "visualization and governance of the entire supply chain," and constitutes an institutional foundation symbolizing the structural evolution of food safety governance in China. As described above, China's food traceability system has been institutionalized in response to serious food safety incidents and, in recent years, has developed toward digitalization and integration. Table 6.2 organizes the development of major policies and institutions in chronological order.

This chronology can also be organized into the following four stages shown in Fig. 6.2.

This developmental process shows that China's food safety governance has gradually shifted from an "ex post response-oriented" model to "institutionalization," and further to a "digital governance-oriented" model. In other words, while the initial stage placed emphasis on responses after the occurrence of food accidents, the system later shifted toward institutional responses through the development of legal institutions and the strengthening of supervisory systems. In recent years, it has developed into an integrated governance model that utilizes digital technologies.

What is particularly important is that traceability is not merely a technical means, but functions as a mechanism that connects legal institutions, administrative supervision, and digital infrastructure in a cross-cutting manner. Through such an integrative function, traceability is positioned as a foundation that supports the assurance of food safety and the formation of trust, and it can be said to have begun to acquire the character of an institutional public good.

6.4.2 The Role of Digital Technologies in Food Safety Issues

6.4.2.1 Blockchain

An important issue in food safety governance is how to ensure transparency and information reliability across the entire supply chain. As discussed in Chap. 1, food passes through multiple stages, including production, processing, storage, logistics, sales, and consumption. Therefore, information discontinuities and uncertainties in quality management are likely to arise at each stage. Particularly in the case of fresh foods and processed foods, it is difficult for consumers to directly verify information concerning the origin of raw materials, the use of pesticides and additives,

Table 6.2 Development of China's food traceability system: chronological overview

Period	Major events and policies	Content	Relationship with traceability
First half of the 2000s	Initial institutional development	Strengthening of supervision based on the Food Hygiene Law (1995)	The concept of traceability remained underdeveloped
2004	Preparations for the Agricultural Product Quality Safety Law began	Strengthening of safety management for agricultural products	Emergence of record management at the production stage
2006	Enforcement of the Agricultural Product Quality Safety Law	Mandatory recording of production history on a trial basis	Institutional starting point for traceability
2008	Melamine incident	Occurrence of a large-scale food safety incident	Trigger for strengthening the traceability system
2009	Food Safety Law, initial version	Legal institutionalization of risk management and supervisory systems	Necessity of traceability became clear
2010–2014	Trial and demonstration stage	Introduction of traceability systems by local governments	Introduction of QR codes and related technologies began
2015	Amended Food Safety Law	Legal clarification of the traceability system	Institutionalization of whole-process traceability
2016	13th Five-Year Plan for Food Safety	Promotion of informatization and digitalization	Aim to build a nationwide traceability system
2017	Proposal for blockchain applications	Discussions on the introduction of IoT and blockchain	Transition to the stage of technological innovation
2018	Establishment of the State Administration for Market Regulation (SAMR)	Integration of the supervisory system	Strengthening of data integration and institutional coordination capacity
2019	Regulations for the Implementation of the Food Safety Law	Strengthening of corporate responsibility and information disclosure	Specification of traceability obligations
Early 2020s	Deepening of digital integration	Utilization of big data and AI	Advancement of traceability

(continued)

Table 6.2 (continued)

Period	Major events and policies	Content	Relationship with traceability
2021	"Digital China" policy	Promotion of data integration and standardization	Orientation toward a nationally unified platform
Since 2022	Strengthening of platform collaboration	Use by Alibaba, JD.com, and others	Integration of private-sector leadership and government coordination

Source Created by the author based on the Standing Committee of the National People's Congress of the People's Republic of China (2009, 2015), the State Council of the People's Republic of China (2019), FAO & WHO (2020), and OECD (2020)

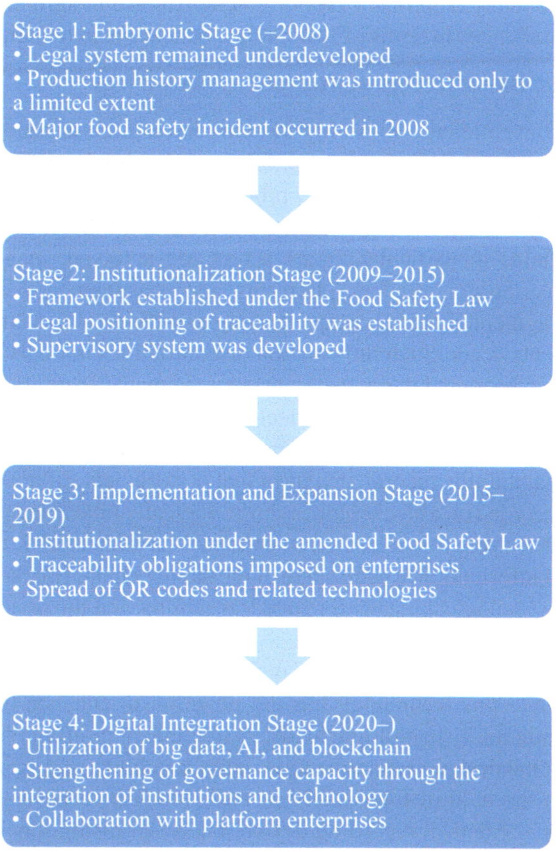

Fig. 6.2 Development of China's food traceability system: four-stage overview (created by the author)

processing procedures, temperature control, and logistics history. For this reason, in food safety governance, how information along the supply chain is recorded, shared, and presented in a verifiable form becomes indispensable for the formation of institutional trust and consumer trust.

In this regard, traceability technologies play an important role as mechanisms that consistently record historical information from the origin of raw materials to final consumption and provide it to relevant actors and consumers. Conventional traceability has mainly been used for internal corporate quality management and administrative supervision. However, with the development of digital technologies, its functions have expanded from mere record management to risk prevention, information disclosure, consumer communication, and even trust formation. In particular, in recent years, digital traceability combining blockchain with visualization technologies such as QR codes has attracted attention as a means of enhancing the transparency of food supply chains. Kamilaris et al. (2019) point out that while there are many empirical cases of blockchain use in agricultural and food supply chains, challenges in terms of technology, education, and institutions constrain its diffusion.

1. Basic Concept of Traceability Technology

Conventional traceability systems have often depended on centralized databases. Such methods have the advantage of clearly identifying the data management actor, but they are also prone to problems such as data tampering, omissions in records, information monopolization, and discontinuities between systems. In food supply chains in particular, many actors are involved, including producers, processors, distributors, retailers, platforms, and consumers. As a result, information is easily fragmented at each stage, making it difficult to quickly identify causes when accidents occur. Such information discontinuities not only expand food safety risks, but also become factors that damage consumers' institutional trust.

In contrast, blockchain technology functions as a distributed ledger and possesses the characteristic of allowing multiple related actors to share the same data while maintaining records in a form that is extremely difficult to tamper with. Information recorded on a blockchain is chronologically linked and preserves a history of changes. Therefore, it is expected to enhance the authenticity, transparency, and traceability of information at each stage of the supply chain. Blockchain can serve as an effective technological means for food traceability and the assurance of authenticity (Galvez et al., 2018).

Moreover, systematic reviews of blockchain applications have shown that while this technology has the potential to enhance transparency, traceability, and auditability in multiple fields, its implementation entails challenges related to institutional design, standardization, interoperability, and governance (Casino et al., 2019). In the field of food safety as well, blockchain needs to be understood not merely as a technical recording device, but as an institutional technology that reorganizes information sharing and responsibility allocation among actors along the supply chain.

Furthermore, blockchain technology can contribute to improving transparency, traceability, and security in sustainable supply chain management, while its introduction requires inter-firm cooperation, institutional support, and standardized data management (Saberi et al., 2019). This point is also important in the field of food safety. In other words, blockchain reduces the cost of information sharing among enterprises and enables the provision of trust to consumers, but for it to be effective, the reliability of data input and institutional supervision are indispensable.

2. Application of Blockchain to Food Traceability

The application of blockchain to food traceability in China has developed in close connection with the advancement of digital governance. In China, legal and institutional development as well as strengthened administrative supervision by the government have progressed in response to food safety incidents. At the same time, the introduction of digital traceability by e-commerce platforms and retail enterprises has also advanced. In particular, in fields such as food e-commerce, fresh foods, imported foods, and high-value-added agricultural products, mechanisms that allow consumers to verify product histories have been emphasized.

A representative case is the introduction of blockchain traceability by Walmart China. Kamath (2018) analyzes Walmart's blockchain demonstration cases for pork and mangoes conducted in cooperation with IBM and evaluates the substantial reduction in the time required for food tracking. In particular, the pork case in China is an important example showing that blockchain can enhance traceability in an environment characterized by a highly complex food supply chain.

In such systems, information from each stage of production, processing, logistics, and retail is recorded, and consumers can verify product histories through QR codes and similar tools. With conventional paper-based records or enterprise-specific fragmented data, identifying causes and confirming the scope of recalls required considerable time when food accidents occurred. However, traceability using blockchain may make it possible to quickly confirm the distribution routes and related lots of target products. This improves the speed of response when food accidents occur and also accelerates the provision of information to consumers.

Within China, various applications by local governments and enterprises have also progressed. In fields such as agricultural products, tea, meat, and imported foods, tracking systems combining blockchain and QR codes have been introduced for the purposes of protecting regional brands, preventing misrepresentation, and ensuring quality. Platform enterprises such as Alibaba and JD.com have also been developing tracking systems that integrate blockchain, IoT, and QR codes, with particular use in imported foods and high-value-added agricultural products. Feng et al. (2020) organize the development methods, advantages, and challenges of introducing blockchain into agri-food traceability, and indicate its potential contribution to improving transparency and forming trust.

In academic research as well, the application of blockchain to food traceability has been highly evaluated based on its technological characteristics. Blockchain has features such as a distributed ledger structure, tamper resistance, and transparency, and it can ensure the consistency and reliability of data at each stage of the supply chain. For this reason, it is considered capable of simultaneously reducing the cost of information sharing among enterprises and improving consumer trust (Galvez et al., 2018; Casino et al., 2019; Kamilaris et al., 2019).

However, despite such technological potential, the application of blockchain to food traceability cannot be said to have become fully widespread at present, either in China or globally. Rather, its stage of development should be evaluated as a transitional phase in which it is moving from demonstration experiments to partial implementation. In fact, although there are cases of adoption by Walmart and some major retail and e-commerce enterprises, these cases are mainly limited to large-scale distribution, high-value-added products, and food categories requiring brand protection. They have not spread comprehensively across entire supply chains that include small and medium-sized enterprises and small-scale farmers.

There are several challenges behind this situation. First is the high cost of introduction. Constructing blockchain traceability requires system development, data input, IoT devices, sensors, QR code management, personnel training, and auditing systems, which impose a heavy burden on small and medium-sized enterprises. Second is the lack of unified data standards. When actors along the supply chain record data in different formats, it is difficult to connect the information and present it as a consistent history. Third is insufficient interoperability among systems. If different traceability systems are constructed by each enterprise, region, or platform, it becomes difficult for consumers and administrative authorities to use the information across systems.

What is even more important is that blockchain does not solve the problem of trust in food safety by itself. Blockchain has the function of preventing tampering with records, but it does not guarantee the accuracy of input data itself. In other words, the problem of "garbage in, garbage out" still remains. If incorrect or false data are entered, even if they are stored in a form that is difficult to tamper with, they do not become reliable information. Therefore, trust cannot be built by blockchain alone if it is separated from on-site inspection systems, third-party certification, administrative supervision, and standardized rules for data input.

For this reason, recent research positions blockchain not as a technology that replaces institutional trust, but as a means of complementing and strengthening it. The introduction of blockchain into supply chain management requires not only the technology itself, but also inter-organizational cooperation, an institutional environment, standardization, and governance structures (Saberi et al., 2019). The same applies to food safety governance. Blockchain provides technological trust, but that trust must be supported by institutional supervision and the responsibility of market actors.

Based on the above, as shown in Table 6.3, although blockchain is a core digital technology supporting the advancement of food traceability, its effectiveness does not depend on the introduction of the technology alone. Rather, it functions only when integrated with institutional design, standardization, administrative supervision, and market incentives. In this sense, trust formation in food safety is a complex process established through the interaction of "technology, institutions, and markets," and the role of blockchain must also be repositioned within this integrated framework.

Table 6.3 Diffusion status of blockchain technology in the food sector

Perspective	Evaluation
Technological maturity	Moderate. Practical application is underway, but full-scale diffusion has not yet been achieved
Degree of diffusion	Limited. It is concentrated in specific fields such as large-scale distribution, high-value-added products, imported foods, and branded foods
Situation in China	Introduction is being promoted relatively actively even by global standards, but diffusion among SMEs and regional distribution channels remains limited
Main challenges	Introduction costs, data standards, interoperability, authenticity of input data, and coordination with supervisory systems
Future prospects	It may expand as part of digital governance, food safety supervision, and e-commerce platform management

6.4.2.2 Integration of Blockchain and QR Code Management

Blockchain greatly contributes to improving the reliability of food traceability. However, in order for consumers to access such information easily, a visualized interface is indispensable. In this respect, QR codes, or Quick Response Codes, play an important role. By scanning the QR code attached to a product package with a smartphone, consumers can access historical information recorded on the blockchain or traceability information provided by enterprises. Such a mechanism is important in that it opens food safety information not only to specialized institutions and administrative bodies, but also to ordinary consumers.

Unlike blockchain itself, QR codes are not a technology that records or guarantees data. Rather, they function as an "entry point" that connects consumers with digital information. If blockchain is a technology that enhances the tamper resistance and shareability of data along the supply chain, QR codes are an interface that visualizes such information for consumers. Therefore, the integration of blockchain and QR codes can be understood as a mechanism that links data reliability at the back end with information accessibility at the front end.

As international examples, traceability systems combining blockchain and two-dimensional codes such as QR codes have been introduced in the field of food distribution. For example, mechanisms that connect product identifiers, lot numbers, expiration dates, distribution histories, and other information to digital information in a standardized form serve as a foundation for consumers and business operators to access product information (GS1, n.d.). Such mechanisms aim to present consumers with information concerning product authenticity, safety, and sustainability, thereby enhancing reliability in food choice. Kamath (2018), through the demonstration case of Walmart and IBM, shows the potential of blockchain to improve the speed and accuracy of food tracking.

In China as well, traceability systems integrating blockchain and QR codes have been introduced across diverse food categories, including local specialty products, imported foods, high-end agricultural products, and fresh foods. In particular, major Chinese e-commerce and retail platform enterprises have constructed mechanisms that present product history information to consumers through QR codes, thereby creating an environment in which consumers can understand information concerning quality and safety before or at the time of purchase. Such mechanisms alleviate the information asymmetry of the traditional situation in which "safety can only be evaluated after purchase," and contribute to the formation of institutional trust based on transparency.

QR codes not only facilitate consumers' access to information, but also have the effect of encouraging enterprises to visualize quality management. If product histories are disclosed to consumers, enterprises must pay greater attention to raw-material management, logistics management, inspection records, and the accuracy of labeling information. In other words, information disclosure through QR codes is not only a service for consumers, but also a mechanism that makes enterprises' internal management externally monitorable. In this respect, QR code management is positioned not merely as a marketing tool, but as part of food safety governance.

Nevertheless, QR code management also has limitations. First, if the information displayed through QR codes is limited, or if enterprises present information selectively, consumers cannot obtain substantive transparency. Second, there is also the possibility that QR codes may be counterfeited or that linked information may not be updated. Third, if consumers cannot understand the displayed information, information disclosure will not sufficiently lead to trust formation. Therefore, in QR code management, the accuracy, updatability, comprehensibility, and standardization of information are important.

6.4.3 The Structure of Food Safety Risks in the E-Commerce Era and an Empirical Analysis of China's E-Commerce Food Market

6.4.3.1 Transformation of Food Safety Governance Through E-Commerce and the Location of the Problem

The rapid expansion of electronic commerce, or EC, has fundamentally restructured food distribution. In China, the online retail market has grown into the largest in the world, and e-commerce transactions involving fresh foods and processed foods have also expanded rapidly. In particular, after the COVID-19 pandemic, online food transactions became established as a routine form of consumption, while O2O, or Online to Offline, delivery models and instant delivery services became generalized. These changes have substantially reshaped the spatial and institutional architecture of food distribution.

Particularly in China, with the expansion of online retail, the modes of food transactions have shifted from face-to-face to non-contact, platform-dependent forms, fundamentally transforming the ways in which food safety risks manifest. According to the National Bureau of Statistics of China, nationwide online retail sales remained at a high level in 2024, and the share of online retail sales of physical goods in total retail sales of consumer goods has been on an upward trend. Moreover, within the online retail sales of physical goods, the food category has shown growth surpassing that of clothing and daily necessities (National Bureau of Statistics of China, 2025). This indicates that the food sector remains a high-growth area even as the overall e-commerce market matures.

Traditional food safety management presupposed physical distribution channels and fixed relationships among business operators, making it relatively clear which actors were subject to supervision and where responsibility lay. Under the e-commerce environment, however, information asymmetry has intensified, driven by the increasing anonymity of transaction actors, the multilayering of supply chains, and the proliferation of platform-mediated indirect transactions. As a result, it has become difficult for consumers to directly verify information concerning product quality and safety, making it necessary to construct the foundation of trust formation in a form different from that of the past. In the OECD Recommendation on Consumer Protection in E-commerce, the OECD points out that cross-border transactions, anonymity, insufficient information disclosure, and difficulties in enforcement are important issues for consumer protection in digital markets (OECD, 2016).

Moreover, as platform enterprises have come to play a central role in food distribution, food safety governance has shifted from mere government regulation to a multilayered governance structure in which platform-based self-regulation, algorithmic management, consumer reviews, and administrative supervision operate in overlapping ways. Such changes expand food safety issues from problems of compliance by individual enterprises into structural issues concerning the institutional design of the entire digital market.

In short, e-commerce does not merely multiply food safety risks; it transforms their character from "localized risks" into "structural risks." Therefore, in addition to conventional supervision-centered governance, integrated institutional responses are indispensable, including the enhancement of information transparency, the clarification of platform responsibility, and the advancement of traceability using digital technologies.

6.4.3.2 Scale, Growth, and Structure of China's E-Commerce Food Market

1. Expansion of Market Scale and Broadening of the User Base

China's electronic commerce market is the largest in the world, and in recent years it has been shifting from a stage of quantitative expansion to one of qualitative transformation. From a population-based perspective, the use of online food services in China has become extremely widespread. According to the 53rd report of the China Internet Network Information Center, or CNNIC, as of December 2023, the number of users of online food delivery had reached approximately 545 million, accounting for 49.9% of all internet users (China Internet Network Information Center, 2024). This scale indicates that online food services have already become established not merely as services limited to some urban consumers, but as a broad social consumption infrastructure.

Furthermore, based on recent trends, it can be considered that the number of users continued to increase in 2025, and that online food services have become established as part of everyday consumption infrastructure. In urban areas in particular, food delivery, fresh food delivery, online grocery services, and instant retail, discussed in Chap. 7, overlap with one another and have greatly eased the temporal and spatial constraints of food distribution. Whereas conventional food purchasing presupposed selection and take-away at physical stores, food consumption in the e-commerce era has come to be constituted as a series of digital processes, including searching on smartphones, comparing prices, checking reviews, placing orders, making payments, and tracking delivery.

Such large-scale and high-frequency use forms an environment in which food transactions occur in real time and at high density. Because food is perishable, freshness-dependent, and requires temperature control, the increase in transaction frequency and the shortening of delivery time enhance consumer convenience while also increasing the difficulty of supervision. In particular, for fresh foods, frozen foods, prepared dishes, dairy products, and similar items, temperature control and time management during the delivery process are directly linked to the maintenance of quality. As a result, food safety management is being compelled to shift from the traditional ex post supervision-oriented model to data-based real-time monitoring-oriented and preventive governance.

As shown in Table 6.4, the total value of online retail sales expanded continuously from approximately 3.9 trillion yuan in 2015 and reached 15.9722 trillion yuan in 2025. Online sales of physical goods alone exceeded 13 trillion yuan, and their share of total retail sales, namely the e-commerce penetration rate, reached 26.1% (National Bureau of Statistics of China, 2026). This trend indicates that the e-commerce market is entering a mature phase after a period of rapid growth. At the same time, the fact that the growth rate of the food category exceeds that of other categories indicates that food e-commerce is still in an expansion phase.

2. Growth Characteristics and Two-Layered Structure of the Food E-Commerce Market.

Meanwhile, even within this process of maturation, the food sector continues to maintain high growth potential. In 2025, the food category recorded a year-on-year increase of 14.5%, greatly exceeding the 1.9% increase for clothing and the 4.1% increase for daily necessities (National Bureau of Statistics of China, 2026). In other words, while China's overall e-commerce market is shifting toward stable growth, a two-layered structure can be observed in which food e-commerce remains in an expansion phase.

This two-layered structure means that food e-commerce is not merely part of the retail channel, but is strengthening its character as living infrastructure. While the e-commerce penetration of clothing, home appliances, and similar categories is

Table 6.4 Integrated table of longitudinal changes in China's e-commerce food market and food safety risks (2015–2025)

Year	E-commerce market size (total online retail sales)	E-commerce penetration rate (physical goods)	Characteristics of market structure	Changes in food distribution	Major food safety risks	Layer of trust
2015	Approx. 3.9 trillion yuan	Approx. 10–12%	Initial stage of e-commerce expansion and mobilization	Food e-commerce remained limited	Quality uncertainty	Psychological trust
2018	Approx. 9.0 trillion yuan	Approx. 18%	Intensification of platform competition	Full-scale growth of food e-commerce	Counterfeiting and mislabeling	Psychological + institutional
2020	Approx. 11.8 trillion yuan	24.9%	Rapid expansion during the COVID-19 pandemic	Rapid growth of fresh food e-commerce	Cold chain	Institutional trust
2021	Approx. 13.1 trillion yuan	26.0%	Concentration around platforms	Daily food e-commerce	Increasingly complex supply chains	Institutional + structural
2022	approx. 13.8 trillion yuan	27.2%	Market maturation	Visible quality differences	Platform responsibility	Structural trust
2023	Approx. 15.4 trillion yuan	27.6%	Strengthening of digital governance	Introduction of traceability	Data reliability	Structural trust
2024	Approx. 15.5 trillion yuan	Approx. 27%	Advanced stage	Deepening of digital integration	Algorithms and information bias	Structural + institutional
2025	15.97 trillion yuan	26.1%	Data-driven market	Food e-commerce as infrastructure	Data governance and trust issues	Structural + institutional

Note The e-commerce market size is based on the "total online retail sales" reported by the National Bureau of Statistics of China, or NBS. This is a statistic based on enterprise reports and does not necessarily correspond to the international concept of sales, but it is the most reliable indicator for understanding trends in China's domestic market. The three-layered trust model (psychological, institutional, and structural) is detailed in Chap. 2

entering a relatively mature stage, food e-commerce remains connected to instant delivery, fresh foods, daily consumer goods, and community-based services, leaving room for further growth. In particular, fresh foods and online groceries are expanding against the background of changing urban lifestyles, the increase in dual-income households, the spread of smartphone payments, and the upgrading of logistics infrastructure.

In terms of the overall market, while China's online retail market maintains one of the world's largest scales, the food and grocery sector is also expanding rapidly. ECDB estimates the sales of China's food e-commerce market at USD 156.264 billion in 2025 and expects growth of 15–20% in 2026 (ECDB, 2025). IMARC Group also presents a high-growth scenario in which China's online grocery market will reach USD 147.2 billion in 2025 and expand to USD 1.0534 trillion by 2034 (IMARC Group, 2025). However, because these private-sector estimates differ in their definitions and coverage depending on the statistical institution, this book positions the online retail statistics of the National Bureau of Statistics of China as the basic data and private-sector estimates as supplementary indicators.

3. Future Forecasts and Growth Drivers

China's online food e-commerce market has expanded rapidly in recent years and has already become one of the core areas within the structure of consumption and distribution. Synthesizing public statistics and industry reports, the market size around 2025 is estimated to have generally reached the scale of one trillion yuan, including food, fresh food, and grocery e-commerce in a broad sense. However, this figure is not a direct estimate from a single institution, but rather a median scenario estimated by the author based on the National Bureau of Statistics' online retail statistics, the growth rate of the food category, and private-sector estimates concerning the online grocery market (Table 6.5).

As shown in Table 6.5, future market trends are expected to shift from the rapid expansion phase observed during the COVID-19 pandemic to a more stable growth stage. From 2026 onward, the growth rate is expected to converge to approximately 8–15% annually, while the market is highly likely to expand to around 1.8–2.1 trillion yuan by around 2030. This can be understood as the result of the simultaneous effects of a declining growth rate accompanying market maturation and the continuing advancement of digitalization, urbanization, and the diffusion of instant delivery

Table 6.5 Forecast of China's online food e-commerce market size after 2026 (median scenario)

Year	Market size	Growth rate	Comment
2025	Approx. 1.05–1.12	–	Based on actual performance
2026	Approx. 1.20–1.30	12–15%	Continued e-commerce penetration
2027	Approx. 1.35–1.48	10–13%	Expansion of instant delivery
2028	Approx. 1.50–1.68	9–12%	Growth in lower-tier cities
2029	Approx. 1.65–1.88	8–11%	Beginning of maturation
2030	Approx. 1.80–2.10	8–10%	Stable growth
2035	Approx. 2.80–3.50	6–9%	Long-term stability

Note Estimated by the author based on the National Bureau of Statistics' online retail statistics, the growth rate of the food category, and market reports including ECDB (2025) and IMARC Group (2025). Therefore, the figures in Table 6.5 are not directly quoted from existing literature, but are estimated based on a median scenario. Unit: trillion yuan (RMB) services. Although some private-sector estimates indicate higher growth rates, this study adopts a more cautious median scenario in light of the maturation of the overall e-commerce market.

The greatest factor supporting this growth is the rapid development of "instant retail." Whereas conventional e-commerce presupposed next-day delivery, platforms capable of delivery within 30 min to one hour are now widespread, and the temporal structure of food distribution itself is being reorganized. The mechanism through which fresh foods, prepared dishes, beverages, frozen foods, and daily necessities are purchased on the same platform and delivered instantly is changing consumer purchasing behavior from "planned purchasing" to "immediate and repetitive purchasing."

The development of e-commerce infrastructure in third- to fifth-tier cities has also become an important driver of market expansion. In regions where the range of physical store options is more limited than in urban areas, e-commerce plays a role in supplementing product variety, price comparison, and delivery convenience. Furthermore, with the spread of mobile payments and regional logistics networks, barriers to online food purchasing are declining even in local cities and rural areas. Therefore, the future growth of the food e-commerce market is considered to be supported not only by demand expansion centered on large cities, but also by the expansion of lower-tier cities and local markets.

4. Market Structure and Structural Transformation of Risk

Behind this market expansion are the spread of digital payments, the upgrading of logistics infrastructure, and the online shift of consumer behavior. In particular, triggered by the COVID-19 pandemic in 2020, the use of e-commerce for daily consumer goods, including fresh foods, spread rapidly, and food e-commerce shifted from a "complementary channel" to a "core distribution infrastructure." Contactless purchasing to avoid infection risks, online ordering under restrictions on going out, and the expansion of home delivery and instant delivery services established new purchasing habits among consumers.

From the perspective of market structure, China's food e-commerce has a typical platform-concentrated structure. Major platforms such as Pinduoduo, JD.com, Alibaba's Tmall, Meituan, and Douyin lead distribution, and they enhance price formation, demand forecasting, delivery efficiency, and sales promotion through vast transaction data and algorithms. On the other hand, this structure also brings about intensified price competition and increased anonymity in transactions, thereby forming new conditions for the emergence of food safety risks.

What is even more important is that changes in market structure and transformations in food safety risks are closely linked. In 2015, the central risks were information-deficiency risks arising from information asymmetry, such as quality uncertainty and insufficient labeling information. However, with the expansion of e-commerce, these risks have shifted toward institutional risks such as mislabeling, quality variation, inadequacies in cold chains, insufficient temperature control during delivery, opacity regarding seller qualifications, and the increasing complexity of supply chains. In recent years, despite the introduction of traceability, new digital-governance-related risks have also become apparent, including the authenticity of data input, the reliability of reviews, information bias caused by algorithms, and manipulation of display rankings by platforms.

This transformation of the risk structure means that food safety issues are expanding from mere quality management issues into issues of the information environment, platform design, data governance, and institutional supervision. In other words, food safety risks in the e-commerce era need to be understood not only as problems of individual products or enterprises, but also as risks inherent in the structure of the digital market itself.

5. Transformation of Trust Structure and Institutional Implications

Such changes in risk structure also correspond to a transformation in the nature of trust in food safety. Conventional trust in food safety was formed on the basis of face-to-face transactions, local relationships, brand recognition, and administrative supervision. However, in the e-commerce era, consumers make purchasing decisions without direct contact with sellers and without confirming the actual products. Therefore, the foundation of trust formation is shifting from psychological trust to institutional trust, and further toward structural trust that depends on platforms and data.

Food is essentially a credence good, and information asymmetry becomes a constraining factor in the market. Akerlof's (1970) discussion of the "market for lemons" clarified that when information concerning quality is unevenly distributed between buyers and sellers, low-quality goods may remain in the market while high-quality goods may exit. In food e-commerce markets as well, when consumers cannot confirm quality or safety in advance, low-quality goods and misrepresented products can more easily enter the market, potentially resulting in a decline in trust in the market as a whole.

However, with the advancement of traceability technologies and data disclosure, consumers are able to access information concerning production, processing, and distribution processes, and the mechanism of trust formation is changing significantly. In particular, the visualization of information using QR codes and blockchain reduces uncertainty regarding food safety and functions as an institutional foundation that supports market expansion. Food safety traceability systems that combine HACCP, Hazard Analysis and Critical Control Point, blockchain, and IoT indicate the potential for tracking information along the supply chain and improving transparency (Tian, 2017).

However, technology does not automatically generate trust. If errors or misconduct occur at the data input stage, even information recorded on a blockchain is not necessarily trustworthy. For this reason, technological transparency becomes effective only when it is linked with institutional supervision, third-party certification, and standardized inspection systems. Therefore, digital technologies in food safety governance should not be positioned as substitutes for institutional trust, but as means of complementing and strengthening it.

In this way, China's online food e-commerce market is shifting from a stage of quantitative expansion to one of qualitative transformation. The focus of market competition is moving from price and convenience toward quality assurance, transparency, and reliability. This change suggests that food safety is being reorganized from a mere object of regulation into an institutional public good supported by digital technologies and institutional design.

China's e-commerce food market is not only globally outstanding in terms of both scale and growth; its development is also accompanied by structural transformations in food safety risks and the reorganization of mechanisms of trust formation. Therefore, when analyzing China's e-commerce food market, it is necessary to understand not only market size and growth rates, but also information asymmetry, platform responsibility, data governance, and structural changes in consumer trust in an integrated manner.

6.4.3.3 Information Asymmetry and Risk Structure in E-Commerce Food Distribution

As described above, the spread of electronic commerce, or EC, has significantly transformed the information structure of food distribution and has expanded information asymmetry more than before. Information asymmetry refers to a condition in which information concerning quality and safety is unevenly distributed among transaction parties, and it is particularly pronounced in experience goods and credence goods such as food (Akerlof, 1970). In traditional distribution, which relied on face-to-face transactions, a certain degree of information was supplemented through appearance, brands, relationships with sellers, and the reputation of physical stores. However, under the EC environment, non-face-to-face and anonymous transactions have become generalized, making it difficult for consumers to directly verify quality in advance.

Under such conditions, the problem of adverse selection, in which low-quality products are more likely to enter the market, becomes apparent, and the structure of food safety risks changes qualitatively. In China's EC food market in particular, multilayered supply chains and platform-mediated transaction structures make information discontinuities likely to arise at each stage of production, processing, and distribution. For example, when producers, wholesalers, logistics operators, platform sellers, and delivery operators exist separately, responsibility for quality management is dispersed across each stage, making it difficult to identify causes and pursue responsibility when problems occur. This creates an environment in which risks such as false labeling, variation in quality, inadequate temperature control, and opacity in expiration-date management are likely to emerge.

Furthermore, the development of the platform economy adds a new dimension to this information asymmetry. In other words, because transaction information is concentrated on platforms and is selected and presented by algorithms, the information that consumers encounter may itself be structurally unevenly distributed. Factors such as search rankings, recommendation displays, the visibility of reviews, and the placement of advertising slots significantly influence consumers' purchasing decisions. However, the mechanisms for presenting this information are not always transparent, and priorities such as sales volume, advertising fees, or price competitiveness may take precedence over quality or safety. As a result, while the volume of information increases, its reliability and transparency are not necessarily guaranteed, creating a situation of "information overload combined with a lack of trust." The OECD has also identified information disclosure in digital markets, the transparency of market participants, and enforcement challenges for consumer protection as key issues (OECD, 2016).

To address this issue, the adoption of traceability technologies and digital certification has been advancing. QR codes, electronic certificates, blockchain, and IoT sensors can visualize the origin of food, processing history, logistics routes, and inspection results, thereby increasing the likelihood that consumers and regulatory authorities can access this information. However, these technologies do not completely eliminate information asymmetry; rather, they primarily change how it is managed. While technologies like blockchain enhance the tamper-resistance of records, they cannot guarantee the accuracy of the input data. Therefore, combining these technologies with institutional oversight, third-party certification, and standardized inspection systems is essential. Casino et al. (2019), in their systematic review of blockchain applications, point out that while blockchain technology has the potential to enhance transparency and traceability, many challenges remain that depend on institutional design and the implementation environment.

Figure 6.3 illustrates how the expansion of information asymmetry accompanying e-commerce triggers a structural evolution of food safety risks, and how, in response, mechanisms for the formation of trust are reorganized.

In traditional vertically integrated distribution, manufacturers and distributors played central roles in quality management. Under the EC environment, however, numerous sellers enter the market, and platforms function as intermediaries, resulting in a tendency for quality responsibility to become dispersed. As a result, differences in the management capacity of individual business operators are likely to manifest directly as food safety risks. In particular, when large numbers of small-scale sellers and regional business operators enter the market, significant variation may arise in compliance with quality standards, storage conditions, labeling rules, and delivery management.

In response, the formation of trust is also reorganized in a three-tiered manner. That is: (1) psychological trust based on consumers' experience and brand perception, (2) institutional trust grounded in legal frameworks and regulations, and (3) structural trust supported by traceability systems and platform governance—all of which are constructed in a mutually complementary way.

Therefore, this figure highlights the need to reconceptualize food safety not merely as a regulatory issue, but as an institutional public good that is established through the integration of institutions, technology, and markets.

Based on the above, food safety risks under the platform economy should not be reduced to the quality management capacity of individual enterprises, but must be understood as "structural issues" inherent in the entire supply chain and platform

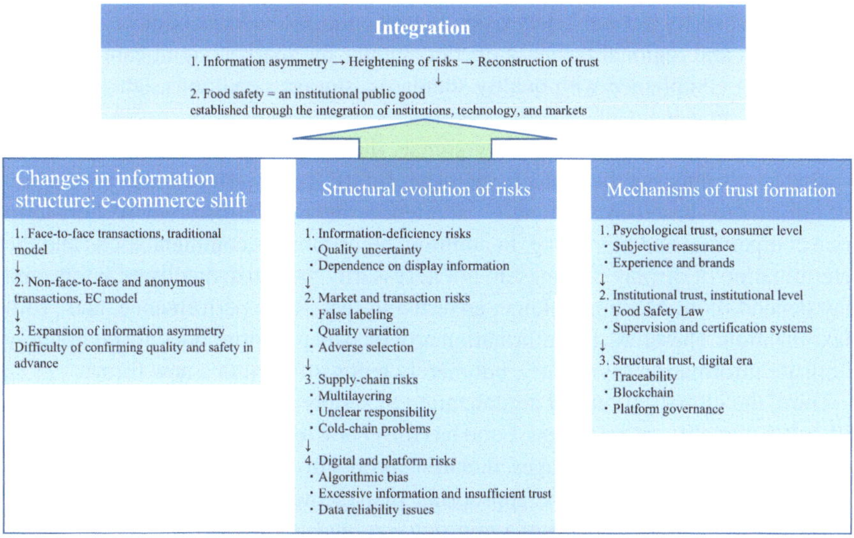

Fig. 6.3 Integration of information asymmetry, risk, and trust in e-commerce food distribution (created by the author)

structure. In this sense, food safety increasingly strengthens its character as an institutional public good supported by the interaction of institutions, technology, and markets, going beyond the issue of mere quality management.

6.4.3.4 Platform Economy and Quality Management Challenges in Supply Chains

With the advancement of electronic commerce, or EC, food distribution is being reorganized into a market structure centered on platforms. The platform economy is characterized by mediating multiple transaction actors—namely producers, distributors, sellers, and consumers—within digital space, and by integrating markets through massive volumes of transaction data and algorithms. Platforms are understood as a new market structure that connects numerous participants and creates value through network effects (Parker et al., 2016). While such a structure dramatically improves distribution efficiency, it also brings new challenges to the way quality management in food safety is conducted.

First, platform-based distribution entails the multilayering and decentralization of supply chains, making it easier for the actors responsible for quality management to become unclear. In traditional vertically integrated distribution, manufacturers and distributors played central roles in quality management. Under the EC environment, however, numerous sellers enter the market, and platforms function as intermediaries, resulting in a tendency for quality responsibility to become dispersed. As a result, differences in the management capacity of individual business operators are likely to manifest directly as food safety risks. In particular, when large numbers of small-scale sellers and regional business operators enter the market, significant variation may arise in compliance with quality standards, storage conditions, labeling rules, and delivery management.

Second, platforms provide quality signals through transaction information and evaluation systems, but their effectiveness has limits. Reviews and ratings are subjective information based on consumers' experiences and may be distorted by short-term price competition and marketing. In addition, algorithmic recommendations and the determination of display rankings do not necessarily prioritize quality or safety, and may depend on transaction volume, advertising fees, sales performance, and profit maximization. Therefore, the information provided by platforms does not completely eliminate information asymmetry, but merely reconstructs it in a new form.

Third, the globalization and digitalization of supply chains increase the technical difficulty of quality management. Food has the characteristic that its quality changes as it passes through multiple stages, including production, processing, transportation, storage, and delivery. Therefore, appropriate management at each stage is indispensable. However, the expansion of rapid delivery and wide-area distribution in EC makes it difficult to maintain cold chains and conduct quality inspections, potentially resulting in quality deterioration and increased safety risks. Kshetri (2018) argues that although blockchain can contribute to supply-chain transparency, accountability, and risk reduction, its introduction requires institutional and organizational conditions.

In order to address these challenges, the introduction of traceability technologies and digital certification has been promoted in recent years. Systems that utilize blockchain and IoT are expected to enhance the visualization and traceability of information across the entire supply chain and to contribute to the strengthening of quality management (Casino et al., 2019). However, these technologies do not solve quality problems by themselves; integration with elements such as institutional supervision, standardization, third-party certification, and platform responsibility is indispensable.

Based on the above, food safety risks under the platform economy should not be reduced to the quality management capacity of individual enterprises, but must be understood as "structural issues" inherent in the entire supply chain and platform structure. In this sense, food safety increasingly strengthens its character as an institutional public good supported by the interaction of institutions, technology, and markets, going beyond the issue of mere quality management.

6.4.4 *Institutional Design of Information Disclosure and Transparency*

6.4.4.1 Information Asymmetry and the Necessity of Institutional Transparency

A central issue in food safety governance is how to reduce the information asymmetry that exists among government, enterprises, and consumers. Food has the characteristics of both experience goods and credence goods, making it difficult for consumers to directly observe its safety and quality in advance. For this reason, the market contains a structural risk of falling into a "market for lemons" caused by quality uncertainty, that is, a phenomenon in which the quality of the market deteriorates due to information asymmetry (Akerlof, 1970). Although Akerlof's discussion originally focused on the used-car market, its theoretical implications are also applicable to food markets, where it is difficult for consumers to directly confirm quality and safety. In particular, in food e-commerce and fresh food distribution, consumers cannot directly observe production processes, processing procedures, storage conditions, or logistics conditions. Therefore, information asymmetry becomes an important factor leading to a decline in market trust.

In China, the 2008 melamine incident greatly shook social trust in food safety, and the reconstruction of institutional trust emerged as a national issue. This incident did not merely reveal illegal conduct by individual enterprises; it also made visible institutional problems concerning the supervisory system, information disclosure, corporate responsibility, and administrative responses. Therefore, restoring trust in food safety required not only the strengthening of penalties and the reorganization of supervisory departments, but also mechanisms through which consumers could access information concerning safety and verify the processes by which institutions operate.

In this context, information disclosure and transparency have come to be positioned not as mere administrative procedures, but as institutional foundations that reduce consumer uncertainty and support trust formation. Transparency corrects the uneven distribution of information in the market and enhances consumers' capacity for choice, while also functioning as external pressure that encourages enterprises to improve quality management and comply with laws and regulations. For administrative authorities as well, information disclosure is an important means of fulfilling accountability and enhancing the legitimacy of regulation. In other words, transparency is not only a means of restoring market efficiency, but also an indispensable precondition for reconstructing social trust.

This point is closely related to the three-layered trust model—psychological, institutional, and structural—presented in this book. Information disclosure first acts on consumers' psychological trust by reducing uncertainty and anxiety. Next, it strengthens institutional trust by visualizing legal systems and administrative supervision. Furthermore, the standardization of information circulation through digital technologies and platforms contributes to the formation of the trust structure of society as a whole, namely structural trust. Therefore, information disclosure and transparency constitute a core mechanism that connects the three layers of psychology, institutions, and structure in food safety governance.

6.4.4.2 Framework for Information Disclosure Based on Legal Institutions

Under this problem awareness, the 2015 amended Food Safety Law of the People's Republic of China clearly positioned the importance of information disclosure in food safety supervision. The law comprehensively stipulates food safety risk monitoring, assessment, standard setting, accident response, supervision and management, and legal responsibility. It reconstructed food safety not merely as an object of ex post administrative sanctions, but as an institutional framework that includes risk management and social supervision (Standing Committee of the National People's Congress, 2015). Under this law, the publication of sampling inspection results, disclosure of administrative penalty information, voluntary information disclosure by enterprises, and food safety risk warning systems were institutionalized, thereby strengthening the framework for correcting information asymmetry.

What is particularly important is that the rapid and accurate publication of food safety risk information was positioned as an important responsibility of supervisory departments. This indicates a shift from the traditional model of internal administrative management to a model based on social participation. In conventional food safety supervision, administrative agencies conducted inspections, supervision, and punishment internally, and in some cases the results were not sufficiently communicated to society. However, through the institutionalization of information disclosure, consumers, media, industry associations, and third-party organizations became able to access food safety information and perform a social monitoring function that complements administrative supervision.

Information disclosure in the field of food safety is also linked to the Regulation on the Disclosure of Government Information. The 2019 amended Regulation on the Disclosure of Government Information of the People's Republic of China clarified the principle that, for the disclosure of government information by administrative agencies, "disclosure is the norm and non-disclosure is the exception," and required information disclosure based on the principles of fairness, impartiality, legality, and convenience for the people (State Council of the People's Republic of China, 2019). Under this institutional framework, food safety information has come to be positioned not as mere internal administrative material, but as information related to the public interest that should be disclosed to society.

On this institutional basis, supervisory agencies, including the State Administration for Market Regulation, or SAMR, continuously publish food safety sampling inspection results, administrative penalty information, consumer warnings, illegal cases, and related materials. As a result, information concerning food safety has increasingly acquired the character of public information shared by society, rather than information closed within administrative agencies. Of course, challenges remain regarding the comprehensiveness of disclosed information, frequency of updates, regional differences, and comprehensibility for consumers. Nevertheless, the institutional expansion of information disclosure itself represents an important change in food safety governance.

Such institutional design directly strengthens "institutional trust" in the three-layered trust model. Institutional trust refers not to trust in individual enterprises or officials, but to the expectation that legal systems, administrative supervision, inspection systems, and information disclosure systems will function stably and fairly. Through the linkage between the Food Safety Law and the Regulation on the Disclosure of Government Information, consumers can access food safety information and confirm the standards by which administrative authorities conduct supervision and what kinds of sanctions are imposed on violating enterprises. This plays a role in enhancing social evaluations of the government's regulatory capacity and accountability.

6.4.4.3 Advancement of Transparency Through Digital Technologies

In recent years, the advancement of digital technologies has brought qualitative changes to the nature of information disclosure. Conventional information disclosure centered on administrative documents, public notices, paper-based reports, or static information posted on websites. However, with the development of the digital economy, information disclosure is shifting toward dynamic mechanisms characterized by real-time availability, searchability, interconnectivity, and verifiability. In the field of food safety as well, databases, online inquiry systems, enterprise credit information disclosure systems, disclosure of food sampling inspection information, blacklist systems, and related mechanisms have been developed, creating an environment in which consumers and social actors can directly access enterprises' violation histories and quality information.

In China, with the development of e-commerce, enormous volumes of data concerning food distribution have come to be accumulated. Online orders, payment information, logistics records, reviews, complaints, returns, delivery times, and temperature-control data can all serve as important sources of information for identifying food safety risks. If these data are appropriately analyzed, it becomes possible to identify at an early stage the concentration of complaints about specific products, abnormal sales patterns, delivery delays, abnormalities in temperature control, and similar issues, and to use such information for risk prevention. Therefore, digital technologies have the potential to expand information disclosure from the mere "publication of results" to "early identification of risks and preventive management."

Furthermore, with the spread of traceability using QR codes, information such as production areas, processing histories, inspection records, logistics routes, and seller information has been visualized, making it possible for consumers to trace the entire life cycle of products. This does not enable consumers to judge product safety entirely on their own, but it does mitigate information asymmetry to a certain extent and strengthens the accountability of enterprises and administrative authorities. Especially in food e-commerce, where consumers make purchase decisions without confirming the actual product, the visualization of digital information plays an important role in trust formation.

In addition, blockchain technology has attracted attention as a means of providing a function for preventing data tampering and enhancing the reliability and verifiability of information. As stated above, food safety traceability systems combining HACCP, blockchain, and IoT indicate the potential for tracking information and improving transparency in food supply chains (Tian, 2017). Blockchain, through its distributed ledger, has the potential to make record tampering difficult and to enhance the reliability of information shared among multiple actors (Tian, 2017). However, blockchain does not guarantee the authenticity of input data itself. Therefore, technological transparency becomes effective only when combined with institutional supervision, third-party certification, and standardized inspection systems.

In this way, through the introduction of digital technologies, transparency is evolving from simple "information disclosure" into an "information environment that anyone can access, compare, and verify." This change is directly connected to the formation of "structural trust" in the three-layered trust model presented in Chap. 2. Structural trust refers not to trust in individual administrative officials or enterprises, but to trust in a social system composed of data, standards, platforms, auditing systems, and technological infrastructure. In contemporary food safety governance, where digital governance is advancing, transparency is becoming a core element that complements institutional trust and forms structural trust.

6.4.4.4 An Integrated Understanding of Transparency and Trust Formation

Nevertheless, transparency itself does not guarantee trust. As risk society theory points out, in modern society, the development of science, technology, and institutions itself generates new risks, and an increase in information does not necessarily lead to an increase in reassurance or trust (Beck, 1992). In food safety as well, the disclosure of large volumes of inspection results, violation information, accident information, and risk warnings may deepen consumers' understanding, but it may also amplify anxiety. In other words, transparency is a necessary condition for trust formation, but not a sufficient condition.

For this reason, institutional trust must be understood as being formed not only through transparency, but also through its interaction with elements such as regulatory enforcement and citizen participation. Even if information is disclosed, if sanctions against violating enterprises are insufficient or enforcement standards differ by region, transparency may instead increase institutional distrust. Moreover, if information disclosure is one-directional and consumers or social organizations cannot participate through opinion expression, reporting, or evaluation, transparency will remain merely administrative publicity and will not sufficiently lead to trust formation.

Therefore, trust should be understood as an integrated concept formed through the interaction of transparency, enforcement, and participation. In this study, this relationship is conceptually expressed as Trust = f(Transparency, Enforcement, Participation). However, this is not a strict mathematical model, but a conceptual expression for organizing the major factors of trust formation. Transparency reduces information asymmetry, enforcement guarantees the effectiveness of institutions, and participation enhances institutional legitimacy and social acceptance. When these three elements complement one another, trust formation across the three layers of psychological trust, institutional trust, and structural trust becomes possible.

From this perspective, the expansion of information disclosure in China can be evaluated as contributing to a certain degree of trust restoration. The amendment of the Food Safety Law, the development of government information disclosure systems, the publication of food sampling inspection information, the disclosure of enterprise credit information, and the introduction of traceability technologies all act in the direction of enhancing transparency. However, their effects differ by region and field, and challenges remain regarding the effectiveness of institutions. In particular, the enforcement capacity of local governments, the quality of data disclosure, the comprehensibility of disclosed information, and enterprises' attitudes toward voluntary disclosure are important factors that determine whether transparency actually leads to trust formation.

Furthermore, when considering the relationship between transparency and trust, it is necessary to distinguish between the "quantity of information" and "reliability." Even if large amounts of information are disclosed, they will not sufficiently contribute to consumer trust formation unless they are accurate, comparable, easy to understand, and timely. Rather, when large volumes of fragmented and specialized information are presented, consumers may find it difficult to make judgments and may consequently experience heightened anxiety and distrust. Therefore, in the institutional design of transparency, it is necessary to create a comprehensive design that includes not only the expansion of the scope of disclosure, but also the quality, format, method of explanation, and usability of information.

6.4.4.5 Institutional Challenges and Structural Constraints

Several structural challenges exist in the current information disclosure system.

First, there is an imbalance in the level of information disclosure among local governments. In China, large differences exist across regions in stages of economic development, administrative capacity, and digital infrastructure. As a result, significant regional differences can be seen in the scope of disclosure of food safety information, frequency of updates, data formats, and searchability. While relatively well-developed online disclosure systems are operated in coastal areas and large cities, in other regions information disclosure may remain formalistic and lack substantive usability. Such regional disparities are a factor that gives rise to spatial imbalance in institutional trust.

Second, there is a "constraint of comprehensibility" caused by the specialized and complex nature of disclosed information. Food safety information includes highly specialized content such as inspection standards, additives, pesticide residues, microbiological indicators, and risk assessment. If such information is disclosed as it is, it is difficult for ordinary consumers to interpret. In other words, transparency is not merely a matter of information quantity; it functions only when accompanied by the comprehensibility and usability of information. In this respect, the current system contains a gap between "disclosure" and "understanding."

Third, there are structural constraints inherent in enterprises' information disclosure behavior. Because the disclosure of quality problems and inspection results affects corporate brand value and market competitiveness, enterprises face misaligned incentives regarding information disclosure. In addition, supply-chain information and raw-material procurement information are closely related to trade secrets, meaning that complete transparency has institutional limits. This indicates that a constant tension exists between public interest and corporate interests in information disclosure systems.

Fourth, there are institutional characteristics from the perspective of international comparison. In the EU and the United States, information sharing and risk communication are incorporated into the core of food safety systems. Preventive control and the institutionalization of information are emphasized in the EU's General Food Law, Regulation (EC) No. 178/2002, and RASFF, as well as in the U.S. Food Safety Modernization Act, or FSMA. In contrast, while China shows a certain degree of convergence with these systems, it is characterized by a governance model linked to digital technologies. In other words, as e-commerce platforms, mobile payments, logistics data, QR codes, credit information disclosure systems, and related mechanisms become interconnected, data-driven and real-time transparency is being formed.

Fifth, there are new challenges associated with the advancement of digital governance. While data integration and analysis are progressing, issues such as algorithmic transparency, data reliability, privacy protection, and uneven distribution of information have become increasingly apparent. In addition, insufficient interoperability among different systems and fragmented data management remain unresolved challenges. These issues suggest that digitalization may enhance transparency while also generating new institutional risks.

As described above, information disclosure systems are built upon complex challenges, including regional disparities, comprehensibility, corporate behavior, institutional design, and new constraints accompanying digitalization. These challenges directly affect the formation of institutional trust and structural trust in food safety, and they indicate the necessity of understanding the function of transparency not as simple "disclosure," but as a more complex institutional process.

6.5 Food Safety as an Institutional Public Good

Food safety has traditionally been conceived as an object of regulation and supervision. With the rise of the digital economy and e-commerce, however, its nature has expanded into a more structural and institutional dimension. This study redefines food safety not merely as a regulatory outcome, but as an "institutional public good" whose benefits extend across society as a whole. This redefinition is based on the classical theory that understands public goods in terms of non-excludability and non-rivalry, while extending it toward a contemporary understanding in which such goods are generated depending on institutional design and governance frameworks. Samuelson (1954) presented the theoretical foundation of public goods and clarified the characteristics of goods that are consumed jointly. Building on this discussion, this study positions food safety not as a public good that exists naturally, but as a public good constructed through the interaction of institutions, technologies, and actors; that is, as an "institutional public good."

Food safety exemplifies this character, being constituted through the mutual interaction of government regulation, corporate quality management, platform-mediated information intermediation, and consumer choices and evaluations. This structure is also consistent with the theoretical framework in which trust functions as a social mechanism for reducing complexity. Trust is understood as a mechanism that reduces social complexity and enables actors to act under uncertain conditions (Luhmann, 1979). In food safety as well, because consumers cannot verify all processes of production and distribution, they must make judgments by relying on institutions, technologies, certification, labeling, and administrative supervision.

Food safety as an institutional public good is composed of three foundations. First is the institutional foundation, consisting of the Food Safety Law, supervisory systems, certification systems, and related mechanisms, which forms the basis of institutional trust. The institutional foundation establishes minimum standards for food safety, deters violations, and clarifies responsibility when accidents occur. Second is the technological foundation, consisting of traceability, blockchain, QR codes, IoT, and related technologies, which ensures information transparency and verifiability. The technological foundation visualizes the processes of food production, processing, logistics, and sales, and facilitates tracking and recall when risks occur. Third is the market and behavioral foundation, represented by e-commerce platforms and consumer behavior, which functions as a feedback mechanism responsible for the formation and reproduction of trust. Consumers' choices, reviews, reports, and evaluations become social pressure that disciplines corporate behavior and strengthens the management responsibilities of platforms.

These three foundations do not function independently of one another, but operate in a mutually complementary manner. Without the institutional foundation, technology would remain a mere information management tool and would have difficulty forming social trust. Without the technological foundation, institutions would remain formal regulations and it would be difficult to visualize the entire actual supply chain. Without the market and behavioral foundation, feedback from consumers and social actors would not be reflected in institutional operation, and governance would remain an administration-centered, one-directional mechanism. Therefore, only through the integration of these three elements can food safety be established as a public benefit extending across society as a whole.

Food safety also has several characteristics as a public good. First, it is non-excludable. Once safety standards and supervisory systems are established, their benefits extend broadly not only to specific consumers, but also to society as a whole. Second, it is non-rivalrous. One consumer's enjoyment of safe food does not undermine the safety of other consumers. Third, it has strong externalities. Food accidents and false labeling may damage trust not only in the relevant enterprise or product, but also in the entire market, the entire industry, and even the national food safety system. Fourth, it is institution-dependent. Food safety does not arise naturally; it depends on institutional design, including legal systems, supervisory systems, standardization, certification, and information disclosure. Fifth, it is dynamic. In response to technological innovation, the development of e-commerce, changes in consumer behavior, and the globalization of supply chains, the risk structure of food safety and institutional responses continue to change.

In particular, the advancement of digitalization accelerates this dynamism and serves as an opportunity to continuously redefine the nature of food safety. The OECD points out that, in consumer protection in digital markets, information disclosure, enforcement cooperation, and responses to cross-border transactions are important (OECD, 2016). This observation also applies to food safety. In food e-commerce, products, transactions, logistics, evaluations, payments, and complaint handling are all interconnected in digital space. Therefore, maintaining food safety requires the integration of data governance and institutional governance.

Furthermore, when integrated with the three-layered trust model presented in this book, food safety can be understood as a public good composed of three layers: psychological trust, institutional trust, and structural trust. Psychological trust refers to the subjective sense of reassurance through which consumers feel that food is safe. Institutional trust refers to the expectation that the Food Safety Law, administrative supervision, inspection systems, certification systems, and information disclosure systems will function stably. Structural trust refers to trust in socio-technical systems, including data, platforms, traceability, algorithms, and third-party audits. In the e-commerce era, the third layer, structural trust, is becoming particularly important, and the maintenance of food safety has come to depend heavily on the quality of digital governance.

From the above perspectives, the institutional development, digital technologies, and e-commerce market analyzed in this chapter correspond respectively to the institutional foundation, technological foundation, and market-behavioral foundation. Institutional development provides the legal and administrative framework supporting food safety. Digital technologies enhance the visualization and verifiability of information and structure transparency. The e-commerce market forms a new arena for trust formation through platforms, consumer evaluations, and transaction data. Through the integration of these three elements, it becomes clear that food safety as an institutional public good is established.

In this sense, food safety is not merely a regulatory issue, but has the character of social infrastructure. If food safety is not ensured, consumers cannot make choices in the market with peace of mind, enterprises cannot engage in sound competition, and administration cannot maintain institutional legitimacy. Conversely, if food safety is stably supplied as an institutional public good, market trust, consumer reassurance, corporate brand value, and the sustainability of society as a whole will be enhanced. Therefore, the maintenance and strengthening of food safety are core issues in institutional design and digital governance.

In conclusion, food safety is a dynamic institutional public good formed through the integration of institutions, technologies, and markets, and its sustainability depends on multi-actor governance and mechanisms of trust formation. In future food safety policy, it will be necessary to promote in an integrated manner not only the strengthening of legal institutions, but also the qualitative improvement of information disclosure, the assurance of the effectiveness of traceability, the clarification of platform responsibility, the institutionalization of consumer participation, and the strengthening of data governance. Through this, food safety can develop not merely as the outcome of administrative regulation, but as an institutional public good shared and maintained by society as a whole.

6.6 Policy Implications: Constructing Trust Governance for Food Safety as an Institutional Public Good

The central policy implication of this chapter lies in the necessity of repositioning food safety not merely as a regulatory object, but as an "institutional public good," and of constructing a governance system centered on trust formation. In particular, China's food safety challenges are rooted in information asymmetry and the fragility of trust, and conventional policy responses centered solely on regulatory tightening have inherent limitations. Future policy design must therefore transition toward a "trust governance" framework that integrates transparency, institutional enforcement, and social participation.

First, transparency must be advanced and institutionalized. Information disclosure is not merely an obligation, but a prerequisite for trust formation. In addition to the publication of sampling inspection results and administrative penalty information, it is necessary to develop databases that are updated in real time and to provide information in forms that consumers can easily access and understand. In particular, the standardization of information disclosure using digital technologies is directly linked to the strengthening of institutional trust.

Second, regulatory enforcement capacity must be strengthened and consistency ensured. In order for information disclosure to be effective, swift and strict enforcement against violations is indispensable. It is necessary to correct disparities in enforcement capacity between the central and local levels, and to promote the integration of supervisory systems and the clarification of responsibility.

Third, digital technologies and traceability must be utilized in an integrated manner. The advancement of traceability using QR codes and blockchain makes it possible to visualize the processes of food production and distribution, and technologically guarantees the reliability of information. Through this, transparency is transformed from formal information disclosure into substantive verifiability.

Fourth, the responsibilities of platform enterprises must be strengthened and co-governance promoted. With the expansion of e-commerce, platform enterprises have become important actors in food safety governance. Alongside the clarification of supervisory responsibility for platforms, it is necessary to construct a framework of collaborative governance among government, enterprises, and consumers.

Fifth, consumer participation and risk communication must be institutionalized. Consumers are not passive actors, but important constituent actors in food safety governance. Through the development of reporting systems and evaluation systems, as well as the promotion of two-way information sharing, consumers' feedback function needs to be institutionally incorporated.

The above policy implications constitute the basic conditions for sustainably providing food safety as an institutional public good. In other words, strengthening psychological trust, institutional trust, and structural trust simultaneously through the integration of transparency, enforcement capacity, and participation is the key to sustainable food safety governance. The progress of institutional reform in China represents an important step in this direction. However, in the future, the construction of a more stable and comprehensive foundation of trust will be required through the qualitative advancement of institutions and the deepening of digital governance.

References

Akerlof, G. A. (1970). The market for "lemons": Quality uncertainty and the market mechanism. *The Quarterly Journal of Economics*, 84(3), 488–500. https://doi.org/10.2307/1879431

Beck, U. (1992). *Risk society: Towards a new modernity*. Sage.

Casino, F., Dasaklis, T. K., & Patsakis, C. (2019). A systematic literature review of blockchain-based applications: Current status, classification and open issues. *Telematics and Informatics*, 36, 55–81. https://doi.org/10.1016/j.tele.2018.11.006

China Food and Drug Administration. (2016). *Measures for the investigation and punishment of illegal acts concerning online food safety* (China Food and Drug Administration Order No. 27). Promulgated July 13, 2016, and effective October 1, 2016.

Codex Alimentarius Commission. (2022). *General principles of food hygiene: CXC 1–1969*. Food and Agriculture Organization of the United Nations & World Health Organization.

Duan, Y., Miao, M., Wang, R., Fu, Z., & Xu, M. (2017). A framework for the successful implementation of food traceability systems in China. *The Information Society, 33*(4), 226–242. https://doi.org/10.1080/01972243.2017.1318325

ECDB. (2025). *Online grocery industry in China 2018–2030*. ECDB.

Feng, H., Wang, X., Duan, Y., Zhang, J., & Zhang, X. (2020). Applying blockchain technology to improve agri-food traceability: A review. *Food Control, 118*, 107365. https://doi.org/10.1016/j.foodcont.2020.107365

Food and Agriculture Organization of the United Nations. (2022). Thinking about the future of food safety: A foresight report. *Food and Agriculture Organization of the United Nations*. https://doi.org/10.4060/cb8667en

Food and Agriculture Organization of the United Nations. (2023). *FAO strategic priorities for food safety within the FAO Strategic Framework 2022–2031*. FAO. https://doi.org/10.4060/cc4040en

Food and Agriculture Organization of the United Nations, & World Health Organization. (2020). *The future of food safety: Transforming knowledge into action for people, economies and the environment: Technical summary of the 2019 international food safety conferences*. World Health Organization.

Frewer, L. (2004). The public and effective risk communication. *Toxicology Letters, 149*(1–3), 391–397. https://doi.org/10.1016/j.toxlet.2003.12.049

Galvez, J. F., Mejuto, J. C., & Simal-Gandara, J. (2018). Future challenges on the use of blockchain for food traceability analysis. Trends in Analytical Chemistry, 107, 222–232. https://doi.org/10.1016/j.trac.2018.08.011

Giddens, A. (1990). *The consequences of modernity*. Stanford University Press.

GS1. (n.d.). *GS1 Digital Link*. Retrieved May 3, 2026, from https://www.gs1.org/standards/gs1-digital-link

IMARC Group. (2025). *China online grocery market: Industry trends, share, size, growth, opportunity and forecast 2026–2034*. IMARC Group.

International Organization for Standardization. (2018). *ISO 22000:2018: Food safety management systems—Requirements for any organization in the food chain*. ISO.

Kamath, R. (2018). Food traceability on blockchain: Walmart's pork and mango pilots with IBM. *The Journal of the British Blockchain Association, 1*(1). https://doi.org/10.31585/jbba-1-1-(10)2018

Kamilaris, A., Fonts, A., & Prenafeta-Boldú, F. X. (2019). The rise of blockchain technology in agriculture and food supply chains. *Trends in Food Science & Technology, 91*, 640–652. https://doi.org/10.1016/j.tifs.2019.07.034

Kshetri, N. (2018). Blockchain's roles in meeting key supply chain management objectives. *International Journal of Information Management, 39*, 80–89. https://doi.org/10.1016/j.ijinfomgt.2017.12.005

Liu, P., & McGuire, W. (2019). Food safety governance in China: Change and continuity. *Food Control, 106*, 106752. https://doi.org/10.1016/j.foodcont.2019.106752

Luhmann, N. (1979). *Trust and power*. Wiley.

National Bureau of Statistics of China. (2025). *Statistical communiqué of the People's Republic of China on the 2024 national economic and social development*. http://www.stats.gov.cn

National Bureau of Statistics of China. (2026). *Statistical communiqué of the People's Republic of China on the 2025 national economic and social development*.

OECD. (2016). *Recommendation of the Council on consumer protection in e-commerce*. OECD Publishing. https://doi.org/10.1787/9789264255258-en

OECD. (2020). *Agricultural policy monitoring and evaluation 2020*. OECD Publishing. https://doi.org/10.1787/928181a8-en

Parker, G. G., Van Alstyne, M. W., & Choudary, S. P. (2016). *Platform revolution*. W. W. Norton.

Saberi, S., Kouhizadeh, M., Sarkis, J., & Shen, L. (2019). Blockchain technology and its relationships to sustainable supply chain management. *International Journal of Production Research, 57*(7), 2117–2135. https://doi.org/10.1080/00207543.2018.1533261

Samuelson, P. A. (1954). The pure theory of public expenditure. *The Review of Economics and Statistics, 36*(4), 387–389. https://doi.org/10.2307/1925895

Standing Committee of the National People's Congress of the People's Republic of China. (2009). *Food safety law of the people's republic of China* (Order of the President of the People's Republic of China No. 9). Adopted February 28, 2009; effective June 1, 2009.

Standing Committee of the National People's Congress of the People's Republic of China. (2015). *Food safety law of the people's republic of China* (2015 revision). Adopted on April 24, 2015, and effective from October 1, 2015.

State Administration for Market Regulation. (2021). *Measures for the supervision and administration of online transactions* (Order No. 37).
State Council of the People's Republic of China. (2019). *Regulation on the disclosure of government information.*
Tang, Q., Li, J., Sun, M., Lv, J., Gai, R. Y., Mei, L., & Xu, L. Z. (2015). Food traceability systems in China. *BioScience Trends, 9*(1), 7–15. https://doi.org/10.5582/bst.2015.01004
Tian, F. (2017). A supply chain traceability system for food safety based on HACCP, blockchain and IoT. In *2017 International Conference on Service Systems and Service Management* (pp. 1–6). IEEE. https://doi.org/10.1109/ICSSSM.2017.7996119
van Dijck, J. (2013). *The culture of connectivity: A critical history of social media.* Oxford University Press. https://doi.org/10.1093/acprof:oso/9780199970773.001.0001
World Health Organization. (2022). *WHO global strategy for food safety 2022–2030.* World Health Organization.
Wu, L., Qin, K., & Chen, X. (2022). Responsibility of citizens in food safety social co-governance in China. *Frontiers in Public Health, 10*, 962629. https://doi.org/10.3389/fpubh.2022.962629
Zhou, J., Jin, Y., & Liang, Q. (2022). Effects of regulatory policy mixes on traceability adoption in wholesale markets: Food safety inspection and information disclosure. *Food Policy, 107*, 102218. https://doi.org/10.1016/j.foodpol.2022.102218

Part III
Digital Economy, Sustainability, and the Reconstruction of Trust: Reconstructing Structural Trust and Sustainable Food Safety Models

Chapter 7
Expansion of China's Online Food Delivery (OFD) Market and New Food Safety Risks: OFD Food Safety in the Era of "Invisible Kitchen"

Weisheng Zhou, Fengyang Wang, and Xuepeng Qian

7.1 Introduction: Why Is OFD a New Food Safety Issue?

In the previous chapter, we analyzed the structural transformation of food safety governance in China from multiple perspectives, including institutional reform, administrative systems, corporate behavior, consumer participation, and the introduction of traceability and digital technologies, to elucidate the mechanisms through which food safety trust is established as an "institutional public good." In particular, we demonstrated a structure in which the tripartite relationship among the government, businesses, and consumers has shifted toward collaborative governance, with trust being built on the foundation of institutional transparency, information reliability, and predictability.

However, the rapid development of the digital economy in recent years has given rise to new food safety risks even within this institutional framework. A prime example of this is the "OFD (Online Food Delivery; hereinafter referred to as OFD)" market, which is the focus of this chapter. The OFD market in China has expanded rapidly against the backdrop of urbanization, an increase in single-person households, long working hours, and the widespread adoption of mobile payments, and has become established as a primary channel for daily meals in urban areas (China Internet Network Information Center, 2023). The market continues to grow, and its

W. Zhou (✉)
College of Policy Science, Ritsumeikan University, Kyoto, Japan
e-mail: zhou@sps.ritsumei.ac.jp

F. Wang
Xi'an Peihua University, Xi'an, China

X. Qian
Graduate school of Global Environmental Studies, Sophia University, Tokyo, Japan
e-mail: qianxp@sophia.ac.jp

W. Zhou (ed.), *Food Safety in China*, https://doi.org/10.1007/978-981-92-3741-8_7

economic impact is immense. This signifies that OFD is transforming from a mere supplementary service into one of the fundamental infrastructures supporting urban life.

The OFD market is characterized by a platform oligopoly centered on Meituan and Ele.me, as discussed in the next section, in which a series of transaction processes, including ordering, payment, store management, delivery, and reviews, are integrated through algorithms. While this platform-driven model significantly reduces transaction costs and promotes the digitalization and market expansion of the food service industry, it also creates structural challenges that differ from traditional food safety governance. Specifically, food safety issues are shifting from problems attributable to individual companies' quality control or violations to "structural risks" stemming from information structures, distribution structures, and labor structures.

First, OFD is premised on the concept of the "invisible kitchen". In traditional restaurants, the dining and cooking space is visible to consumers, and a certain degree of social oversight functions. However, with OFD, consumers cannot directly verify the cooking environment or hygiene conditions and are forced to rely on information and reviews on the platform. As a result, information asymmetry increases, making it difficult to visualize the actual state of hygiene management. In particular, it has been pointed out that safety issues in online food distribution are strongly rooted in the supply structure itself mediated by the platform (Li et al., 2020).

Second, OFD introduces a new step, "delivery", between preparation and consumption. Inadequate temperature control during delivery, insufficient container sealing, quality deterioration due to long transport times, and the risk of secondary contamination are issues that rarely surfaced in traditional dine-in settings. Because consumers cannot directly observe these risks, a disconnect between subjective perception and reality is likely to occur, potentially amplifying risk perception (Kasperson et al., 1988).

Third, the OFD market is closely linked to the gig economy. While delivery personnel are offered flexible employment opportunities, the structure of this system, characterized by labor mobility and the difficulty of standardization, can impact service quality and safety management. In such an environment, it is well established that consumer trust in food safety depends not only on objective safety but also heavily on subjective risk perception.

As described above, OFD is not merely the digitization of the existing food service industry; rather, it is accompanied by a new institutional environment characterized by: (1) spatial invisibility, (2) the multi-layered nature of the supply process, (3) the redistribution of responsibility through the platform, and (4) algorithmic management. This shift transforms traditional food safety issues from "individual risks" into "structural risks," which cannot be adequately explained solely as a matter of psychological trust.

Therefore, in this chapter, building on the analytical framework of institutional trust and collaborative governance presented in Chap. 6, we examine how food safety risks are being reconfigured in the OFD market. In particular, we focus on governance mechanisms unique to the digital age, such as the role of platform companies, algorithmic management, and consumer reviews, to clarify how food safety governance is expanding and transforming. OFD is a phenomenon that symbolizes the shift in China's food safety issues from "manufacturing- and store-centered safety" to "platform-based distribution safety," and it offers a new analytical perspective for food safety research.

7.2 The Rapid Expansion of the OFD Market and Industrial Structure in China

7.2.1 The Rapid Growth and Structural Characteristics of the OFD Market

7.2.1.1 Overview: Transitioning to an "Infrastructure Industry" over the Next 10 Years

China's OFD market has expanded rapidly over the past decade and has grown into one of the world's largest digital service industries. According to the China Internet Network Information Center (CNNIC), the number of OFD users reached approximately 545 million as of 2023, accounting for about half of all internet users (China Internet Network Information Center, 2023). This surge represents a dramatic expansion from the tens of millions of users in the mid-2010s, indicating that OFD has penetrated not only urban areas but also provincial cities and suburban regions, transforming into a primary channel for daily food consumption.

Table 7.1 illustrates how China's OFD market has expanded in scale over the past decade and transformed industrial and social structures. The key points lie not merely in growth, but in "phased development" and "structural transformation." As shown in Table 7.1, China's OFD market experienced explosive growth between 2015 and 2025. The market size expanded from approximately 100 billion yuan to about 1.6 trillion yuan, marking a more than tenfold increase. Additionally, the number of users rose from approximately 100 million to between 550 million and 600 million, reaching a level approaching half the population. Furthermore, the number of delivery personnel has expanded from approximately 1.5 million to ten million. As a result of this quantitative expansion, it can be said that OFD has transformed from a supplementary service into a fundamental infrastructure supporting urban life.

7.2.1.2 Growth and Phased Structure of the OFD Market

Based on existing statistics and related research, the development of the OFD market can be categorized into the following three stages in this study.

Table 7.1 Changes in China's food delivery market over time (2015–2025)

Year	Number of users (100 million people)	Market size (100 million yuan)	Number of transactions (100 million orders)	Delivery staffs and related employees (10,000 people)	Notes
2015	Approx. 1.1	Approx. 1200–1500	Approx. 20	Approx. 150	Market formation phase
2016	Approx. 2.1	Approx. 1700–2000	Approx. 30	Approx. 200	Expansion of O2O competition
2017	Approx. 3.4	Approx. 2400–3000	Approx. 45	Approx. 300	Rapid increase in users
2018	Approx. 4.1	Approx. 3500–4500	Approx. 60	Approx. 400	Market oligopolisation centered on Meituan and Ele.Me
2019	Approx. 4.2–4.6	Approx. 6000	Approx. 80	Approx. 500	Integration into urban life
2020	Approx. 4.2	Approx. 8000–8500	Approx. 100	Approx. 600	Expansion of contactless demand during the COVID-19 pandemic
2021	Approx. 5.44	Over 10,000	Approx. 120	Approx. 650–700	Online food delivery becomes part of daily-life infrastructure
2022	Approx. 5.22	Approx. 10,500	Approx. 130	Approx. 700	Maturation and strengthening of regulation
2023	Approx. 5.45	Approx. 12,000	Approx. 150	Over 1000	Expansion of instant-delivery labor
2024	Approx. 5.92	Approx. 15,000	Approx. 160	Over 1000	Users account for 53.4% of internet users
2025	Approx. 5.69	Approx. 16,000	Approx. 160–180	1000–1200	Intensified competition and expanded entry by JD.com and Taobao-related platforms

Note Compiled and estimated by the author based on multiple statistical sources and existing research regarding China's OFD market. User numbers are based on statistics from the CNNIC (China Internet Network Information Center, 2023), while market structure and service characteristics were derived from existing research (Li et al., 2020). Additionally, government-published statistics were consulted as a supplementary source for overall trends regarding the number of employees (National Bureau of Statistics of China, 2024). Due to differences in statistical definitions, each indicator should be interpreted as an approximate figure

1. **Market formation phase (2015–2017)**

During this period, the market was taking shape alongside the establishment of the O2O (Online to Offline) model. Against the backdrop of widespread smartphone adoption and the expansion of mobile payments, the convenience of OFD services was rapidly recognized, and the number of users surged from approximately 100 million to 300 million. At this phase, OFD began to be accepted as a "convenient service."

2. **Expansion phase (2018–2020)**

Subsequently, the OFD market entered a period of expansion, giving rise to a platform oligopoly dominated by Meituan and Ele.me. Delivery networks were established on a national scale, significantly improving service capacity. Furthermore, the spread of COVID-19 in 2020 led to a surge in demand for contactless services, marking a turning point that transformed OFD into an "essential service" (China Internet Network Information Center, 2023).

3. **Maturity phase (2021–2025)**

Since 2021, the sector has entered a mature phase, with the number of users stabilizing at over 500 million and the market size exceeding 1 trillion yuan. At the same time, the number of delivery personnel has expanded to around ten million, reinforcing the role of OFD as both a "social infrastructure" and a "massive employment market" (National Bureau of Statistics of China, 2024).

7.2.1.3 Key Structural Characteristics of OFD

To summarize the above, China's OFD market exhibits the following structural characteristics.

First, the number of users has already reached half the population, and growth appears to be plateauing. As a result, the market is shifting toward a structure where future expansion will be driven not by an increase in new users, but by higher usage frequency and greater reliance on the service (China Internet Network Information Center, 2023).

Second, the number of transactions has reached 10 billion annually, with tens of millions of transactions occurring daily at an extremely high frequency, transforming food distribution into a real-time, high-frequency process. This creates a temporal structure distinct from traditional food service, signifying the formation of an instant distribution system where supply and demand are connected instantly. These characteristics have also been noted in existing research as evidence that OFD platforms function as high-frequency, highly immediate services (Li et al., 2020).

Third, the OFD market has created a gig economy comprising approximately ten million delivery workers. This expansion has had a significant impact on the labor market, giving rise to a new form of employment characterized by low barriers to entry, high mobility, and algorithm-driven labor management. While OFD work provides flexible employment opportunities and serves as a source of job creation, it also entails unstable working conditions and safety risks due to algorithm-driven time management and competitive pressures (Sun et al., 2021). Furthermore, this high labor turnover makes it difficult to standardize food handling procedures and ensure thorough training, which in turn affects food safety management. Fig. 7.1 shows OFD delivery workers resting and waiting on a street in Shanghai.

Fig. 7.1 OFD delivery workers resting and waiting on a street in Shanghai (Photographed by the Author)

7.2.1.4 Forecast for the Future Chinese OFD Market Size

Table 7.2 presents representative forecast ranges for the Chinese OFD market, compiled based on official statistics and existing research. The Chinese OFD market is projected to reach approximately 1.6 trillion yuan by 2025, expand to around 1.9 trillion yuan by 2027, and exceed 2 trillion yuan by 2030. While the annual growth rate is converging to around 5–10%, stable growth is still anticipated. As mentioned earlier, this indicates that the market is transitioning from a rapid growth phase to a mature phase; going forward, growth will be driven not by an increase in the number of users, but by higher usage frequency and more sophisticated services (China Internet Network Information Center, 2023; National Bureau of Statistics of China, 2024).

Table 7.2 presents a forecast of the growth trajectory of China's OFD market from 2025 to 2030, clearly illustrating the market's transition from a rapid growth phase to a mature phase. First, the market size in 2025 is estimated to be approximately 1.6 trillion yuan, which serves as the base year based on actual performance. At this stage, the OFD market has already reached a massive industrial scale and has become established as essential infrastructure for urban life. In 2026, the market is expected

Table 7.2 Forecast of the scale of China's food delivery market (2026–2030)

Year	Market size (trillion yuan, estimate)	Notes
2025	Approx. 1.6 (base year)	Based on actual performance
2026	Approx. 1.7	Mature growth phase
2027	Approx. 1.95	Consistent across multiple surveys
2028	Approx. 2.1	Moderate expansion
2029	Approx. 2.2	Stabilization of growth
2030	Approx. 2.3	Mature market

Note This table presents the author's compilation and estimates regarding future projections for China's OFD market, based on official statistics and existing research. Future market size figures are estimated based on internet usage trends and the development trends of the digital services market (China Internet Network Information Center, 2023; National Bureau of Statistics of China, 2024; Li et al., 2020), and each figure should be interpreted as an approximation due to differences in statistical definitions and market scope

to expand to around 1.7 trillion yuan, entering a "mature growth phase." This stage is characterized by stable growth rather than rapid expansion. It is projected to reach approximately 1.9 trillion yuan in 2027, a level that can be considered a benchmark for mid-term market expansion.

Furthermore, the market is expected to expand gradually to approximately 2.0–2.1 trillion yuan in 2028 and approximately 2.1–2.2 trillion yuan in 2029, transitioning into a phase of stable growth. During this period, the primary drivers of growth will be increased usage frequency and service sophistication rather than an increase in the number of users. By 2030, the market is expected to reach approximately 2.2–2.3 trillion yuan, at which point it is considered to have clearly entered a mature stage. At this stage, OFD will transcend the scope of a mere food delivery service and function as a broad-based lifestyle infrastructure that includes "instant retail."

To summarize the above, the future outlook presented in Table 7.2 is as follows: first, while the market size itself will continue to expand, its growth rate will gradually decline; second, the market's growth mechanism will shift from "increased user numbers" to "deeper usage"; and third, the nature of the industry will change from a rapidly growing business to a stable infrastructure industry. Therefore, it can be concluded that China's OFD market will move beyond the stage of quantitative expansion and transition into a mature market where qualitative advancement and institutional development become the central challenges.

7.2.1.5 The Relationship Between the Rapid Growth of the OFD Market and Food Safety

The OFD market has expanded rapidly worldwide in recent years and is becoming an established part of daily life, particularly in urban areas centered on China. While this market expansion has improved consumer convenience and accelerated the digitalization of the foodservice industry, it has also significantly altered the landscape of food safety risks.

First, the rapid growth of the OFD market has led to the decentralization of the food supply process. Whereas traditional food service completed the cycle of "preparation-service-consumption" within a single space, OFD spatially separates "preparation-delivery-consumption." Due to this structural change, food safety risks are no longer confined to the kitchen but extend across the entire supply chain, including delivery and storage. In fact, research on online food distribution has pointed out that food safety risks exist across multiple stages, including production and processing, distribution and delivery, and consumption (Li et al., 2020).

Next, the rapid expansion of the market has led to an explosive increase in the number of entities subject to oversight, widening the gap between regulatory capacity and demand. The explosion of food delivery platforms has lowered the barriers to entry for restaurants into the online market, resulting in a rapid increase in the number of suppliers. Consequently, the situation has arisen where traditional on-site inspections and individual supervision by government agencies are no longer sufficient. In this regard, research on governance in the platform economy points out that the complexity of supervision increases alongside market expansion (Calo & Rosenblat, 2017).

Furthermore, the growth of OFD is altering consumer behavior itself and giving rise to new aspects of food safety risks. It has been noted that the increase in online orders is linked to more frequent dining out and greater access to high-calorie foods, which is associated with potential increases in health risks and foodborne illnesses (Jia, 2020). In this way, OFD is not merely a means of distribution but has become a structural factor that transforms dietary patterns and patterns of risk exposure themselves.

Furthermore, intensifying market competition is another key factor affecting food safety. Increased price competition and the race to offer faster delivery times may lead to a decline in hygiene and quality control due to pressure to cut costs. Recent regulatory actions by authorities have signaled a trend toward requiring platform companies to strengthen vendor screening and safety management, suggesting a risk that rapid growth and competitive pressure could weaken safety management systems.

In summary, the rapid growth of the OFD market is reshaping food safety issues through four key aspects: (1) the multi-layered nature of risks resulting from supply chain fragmentation, (2) a mismatch with regulatory oversight capacity, (3) changing consumer behavior, and (4) the risk of quality deterioration due to competitive pressures. Therefore, food safety in the OFD sector must be viewed not merely as a matter of strengthening hygiene management, but as a challenge requiring integrated governance that includes platform regulation, risk-based supervision, and data-driven visualization.

7.2.2 *Platform Oligopoly Structure*

7.2.2.1 A Duopoly Dominated by Meituan and Ele.Me

China's food delivery market has undergone intense competition amid rapid expansion, but it has now effectively consolidated into an oligopoly dominated by two major platforms. At the core of this market are Meituan (Meituan, 2024) and Ele.me (Alibaba Group, 2024).

First, Meituan operates as a "comprehensive lifestyle services platform," integrating not only food delivery but also hotel reservations, travel, leisure, and group buying, with its food delivery division serving as its core business. Leveraging a large-scale delivery network and algorithmic management, Meituan has highly integrated pricing, store ranking, and delivery route optimization. This integrated platform structure not only dramatically improves the efficiency of matching supply and demand but also serves as a factor in strengthening market dominance through network externalities. Its market share is estimated to be around 60%, establishing a dominant position in the market (Meituan, 2024).

Meanwhile, Ele.me is backed by Alibaba-affiliated capital and leverages its integration with e-commerce platforms as a key strength. Its integration with Alipay's payment infrastructure and e-commerce data serves as part of a customer retention strategy, enabling data-driven marketing and integrated logistics. Its market share is estimated to be around 30%, making it the second-largest player behind Meituan (Alibaba Group, 2024).

Table 7.3 presents a time-series analysis of the market share trends for Meituan and Ele.me, the leading platforms in China's OFD market, from 2015 to 2025, clearly illustrating the process by which the market structure is converging from a competitive market to an oligopolistic (duopoly) market.

1. **Subsidy competition period (2015–2017)**

First, from 2015 to 2017, the two major platforms in China's food delivery market, Meituan and Ele.me, implemented large-scale price subsidy policies aimed at acquiring users and partner restaurants. This period represented a classic phase of "subsidy competition," during which platform companies adopted strategies that prioritized expanding market share and network scale even at the expense of short-term profitability (OECD, 2020). Specifically, coupons and discounts were widely offered to consumers, while merchants received reduced or waived fees and sales promotion support. As a result, usage frequency increased rapidly, and the market as a whole expanded explosively.

On the other hand, since this kind of subsidy competition relies heavily on capital strength, companies with superior fundraising and operational capabilities tended to gain a stronger advantage. During this process, Meituan expanded its market share by leveraging its service integration capabilities and economies of scale, gradually widening the gap with Ele.me. As a result, the competitive landscape gradually shifted

Table 7.3 China's food delivery market: market share and HHI trends of Meituan vs. Ele.me (2015–2025)

Year	Meituan (%)	Ele.me (%)	Others (%)	HHI	Characteristics
2015	50	40	10	4200	Early stage of competition; subsidy-driven competition period
2016	55	35	10	4350	Oligopolization progressed after mergers and integration
2017	60	30	10	4600	Formation of a duopolistic structure
2018	63	31	6	4966	After Alibaba's acquisition of Ele.Me
2019	65	30	5	5150	Stable duopolistic structure
2020	68	27	5	5378	Expansion of Meituan's dominance
2021	**69**	28	3	5554	Expansion of demand during the COVID-19 period
2022	67	28	5	5298	Market maturation period
2023	67	28	5	5298	Structure almost fixed
2024	65	33	2	5318	Securities company data
2025 (first half)	60	33	7	4738	Industry reports
2025 (second half)	49	38	13	4014	Reignition of competition; structural change

Source Compiled by the author from Meituan (2024) and Alibaba Group (2024), China Internet Network Information Center (2023), OECD (2020). *Note* Market shares are indicative estimates

toward a dominant player, and it can be said that the foundation for the subsequent oligopoly was laid during this period.

2. **Formation of an oligopolistic market structure (2018)**

Next, 2018 marked a significant turning point. Triggered by Alibaba's acquisition of Ele.me, the market underwent a restructuring, and an oligopolistic structure dominated by the two major platforms became clear. This consolidation went beyond a simple corporate acquisition; it signified the integration of ecosystems encompassing payment, logistics, and data infrastructure, shifting competition between platforms to a more sophisticated form of integrated competition.

From this point onward, a structure emerged in which Meituan maintained a market share of around 60% and Ele.me around 30%, and the market transitioned to a stable duopoly (Alibaba Group, 2024). This demonstrates the characteristics of a typical platform market, where network externalities and switching costs make it difficult for new entrants to enter.

3. **Period of stable two-party dominance (2019–2024)**

Furthermore, while demand for OFD surged between 2020 and 2022 due to the impact of the COVID-19 pandemic, market share itself did not fluctuate significantly; rather, the oligopolistic structure was reinforced. During this period, a structure in which the two companies accounted for the majority of the market became established, and the Chinese OFD market has been functioning as a typical duopoly (China Internet Network Information Center, 2023).

This stable oligopolistic structure is underpinned by factors such as economies of scale, network externalities, and data accumulation; it is believed that the interaction of these factors has solidified their competitive advantage. It is also important to note that competition has shifted from traditional price competition to "data competition," with algorithm-driven pricing and the optimization of delivery efficiency becoming central to the competitive landscape.

4. **Re-competition period (2025 and beyond)**

However, in recent years, signs of change in the market structure have emerged due to new entrants and renewed price competition. Fluctuations in market share have been observed, and it has been pointed out that the previously stable oligopolistic structure may be partially destabilized. In particular, it is well known that competition in digital platform markets is often reignited by technological innovation and regulatory changes, and the Chinese OFD market is no exception.

These developments suggest a cyclical dynamic of competition and concentration in the platform economy, which can be understood in the long term as a process of structural change characterized by "competition-concentration-re-competition" (OECD, 2020).

7.2.2.2 Characteristics of Oligopoly

1. **Market concentration**

Regarding the Chinese food delivery market (Meituan vs. Ele.me), an analysis of market concentration using the Herfindahl–Hirschman Index (HHI) based on the market shares shown in Table 7.3 yields the following assessment.

First, according to the U.S. Department of Justice (DOJ) criteria, a market with an HHI below 1500 is classified as competitive, a market with an HHI between 1500 and 2500 is classified as moderately concentrated, and a market with an HHI exceeding

2500 is classified as highly concentrated (U.S. Department of Justice, 2010). Based on these criteria, it is estimated that China's OFD market has consistently maintained an HHI well above 4000 since 2015, confirming that it is a market with an extremely high degree of concentration.

In particular, around 2020, the HHI is estimated to have reached approximately 5378, marking the peak of market concentration. This level indicates a de facto duopoly structure that goes beyond mere high concentration, marking a stage where market dominance by the two major platforms, Meituan and Ele.me, was firmly established. Furthermore, this high concentration can be interpreted as the result of rapid market consolidation driven by capital strength and network externalities following the subsidy competition period from 2015 to 2017.

Subsequently, while the HHI is expected to decline slightly to approximately 5298 in 2022, it is believed to maintain an extremely high level of concentration. Furthermore, given that this high level is projected to continue beyond 2024, the OFD market can be assessed as having transitioned to a stable oligopolistic structure.

However, it has been pointed out that the HHI may decline to around 4000 in 2025, indicating signs of renewed competition driven by new entrants and shifts in competitive strategies. Such fluctuations reflect the dynamic characteristics unique to platform markets, namely, a structure in which concentration driven by network externalities and dispersion driven by regulation, technological innovation, and strategic competition interact cyclically (OECD, 2020).

In summary, China's OFD market can be understood as a dynamic oligopoly accompanied by a resurgence of competition in the short term, while maintaining the characteristics of a highly concentrated market in the long term. This suggests that market structures in the platform economy cannot be captured by a static dichotomy of monopoly and competition, but rather are characterized by the "cyclical coexistence of concentration and competition."

2. **The institutional impact of market oligopoly**

The oligopoly formed by these two companies does not merely signify market concentration; it also has systemic implications. First, the platforms hold significant bargaining power to determine whether restaurants can list their businesses, the commission rates they pay, and their visibility. For small and medium-sized restaurants that rely on OFD sales, these platforms are effectively indispensable infrastructure, and there is a tendency for pressure to intensify regarding price competition and the burden of promotional expenses.

Second, delivery workers' working conditions and evaluation criteria are also managed by algorithms, which significantly impacts the structure of the labor market. In particular, research on platform work points out that while algorithmic management increases labor flexibility, it also leads to greater instability and risk in working conditions (Sun et al., 2021).

Furthermore, the oligopolistic structure is significant from the perspective of food safety governance. While the traditional food service industry relied primarily on direct government oversight of establishments, in the OFD market, platforms function as de facto "quasi-regulatory bodies." In other words, the vetting of establishment information, the online display of business licenses, and the removal of non-compliant establishments are all managed according to the platforms' internal rules.

However, the transparency of rating algorithms and review criteria is limited, and concerns have been raised that this could lead to ambiguity regarding accountability. This issue is being debated internationally as a governance problem for digital platforms, and striking a balance between regulation and self-governance has become a critical challenge (OECD, 2020).

Thus, China's OFD market has converged into an oligopolistic structure centered on two major platforms. While this has reduced transaction costs and realized economies of scale, it has also led to the concentration of market power and a realignment of responsibility. Food safety issues are no longer merely a matter of individual store management; they must be understood as structural challenges closely linked to platform design, algorithmic governance, and commission structures.

7.2.3 Delivery Labor and the Gig Economy

7.2.3.1 Flexible Employment under Algorithm Control and its Structural Challenges

The rapid expansion of China's OFD market has been accompanied by the formation of a large-scale delivery workforce. OFD delivery workers are typically classified as gig workers who accept jobs through platforms; they generally adopt a flexible work arrangement based on piecework pay rather than fixed employment contracts. This form of employment has become a significant source of jobs for rural migrants and young people moving to urban areas, and the OFD industry functions as a buffer for the urban labor market (Sun et al., 2021).

Major platforms such as Meituan and Ele.me are said to have built large-scale delivery networks, and the operation of OFD services is supported by a large number of gig workers (China Internet Network Information Center, 2023). Delivery workers accept orders displayed on the app and make deliveries following the optimal route calculated by an algorithm. During this process, metrics such as order acceptance rate, delivery time, and customer ratings are recorded in real time and reflected in their compensation and the allocation of future orders. In other words, labor management is shifting from human supervision to "digital governance" mediated by data and algorithms (Sun et al., 2021).

While this gig economy structure offers the advantage of employment flexibility, it also entails several structural challenges. First, piece-rate pay creates an incentive to maximize the number of deliveries, thereby increasing time pressure. It has been pointed out that this may lead to traffic violations and a disregard for safety. Second, institutional gaps in worker protection, such as inadequate social insurance coverage and ambiguities regarding workers' compensation, have become problematic (International Labour Organization, 2021). Third, the high turnover rate among delivery workers makes it difficult to consistently provide standard education on food handling and hygiene training, which can undermine the consistency of food safety management.

A key point regarding food safety is that the delivery process has become an integral part of the food supply chain. In traditional dine-in settings, the time from preparation to consumption is relatively short, and temperature and hygiene control were primarily managed within the restaurant. However, in OFD, the transportation process handled by delivery personnel directly impacts quality maintenance. Delays in delivery, damaged containers, and insufficient temperature-control equipment can lead to secondary contamination and quality deterioration. Such risks have been identified in existing research on online food delivery as food safety challenges that extend across the entire supply chain (Li et al., 2020).

Thus, while the gig work structure in OFD contributes to job creation through labor market flexibility, it also creates new challenges in both safety management and labor protection under the competitive pressures brought about by algorithmic management and piecework-based compensation. Food safety issues must now be understood as structural problems that depend not only on the kitchen environment but also on the institutional design of the entire "extended supply chain," including delivery.

7.2.3.2 Composition of OFD Market Delivery Personnel

As mentioned earlier, the number of workers engaged in delivery services, primarily on major platforms such as Meituan and Ele.me, is estimated to be around ten million, demonstrating an unprecedented capacity to absorb labor for a single service industry. This scale indicates that the advancement of China's digital economy goes beyond merely enhancing consumer services and plays a significant role in creating large-scale employment (Meituan, 2024).

Looking at the demographic composition of delivery workers, a clear skew in the age distribution is evident. The average age is generally around 30, with the 21- to 40-year-old age group constituting the core workforce and accounting for the majority of the total. In particular, the group spanning the early to late 30 s is substantial, representing an age range where physical requirements and the need to secure income overlap. On the other hand, the proportion of those under 20 and over 50 is low, confirming that the OFD delivery workforce is a labor market centered on young and middle-aged adults. Furthermore, men constitute the majority of the workforce, a trend likely driven by the nature of the work, which involves long hours, physical strain, and traffic risks (Sun et al., 2021).

Regarding educational background, while systematic statistics are limited, a synthesis of multiple surveys indicates that workers with high school or secondary vocational education constitute the majority, although it has been noted that the number of college and university graduates entering the sector has been increasing in recent years. Particularly in urban areas, against a backdrop of intensifying competition for jobs and employment insecurity among young people, a certain proportion of highly educated individuals tend to engage in delivery work on a temporary or part-time basis. According to a report by the International Labor Organization, platform work is noted to serve not only as a receptacle for low-skilled labor but also as a safety valve for the labor market (International Labor Organization, 2021).

Furthermore, it is important to note that the employment status of delivery workers is extremely fluid. Only a small proportion of workers are engaged in full-time, continuous employment; most opt for irregular working arrangements involving short-term or intermittent participation. In reality, only a minority of delivery workers maintain high activity levels throughout the year; most view this work as a side job or a temporary source of income. This high fluidity is a characteristic phenomenon of platform work, and, combined with algorithm-driven supply-demand adjustments, it increases the flexibility of the labor supply (China Internet Network Information Center, 2023).

In light of the above, China's OFD delivery worker labor market exhibits three key characteristics: (1) a massive employment capacity of approximately ten million people, (2) a workforce composed primarily of young people, and (3) an unstable and fluid gig-work model governed by algorithms. This structure functions not merely as a delivery service but as an "employment buffer" within the digital economy.

At the same time, this gig work structure is closely linked to food safety. Specifically, while delivery workers serve as the "last mile" connecting restaurants and consumers, contributing to shorter delivery times and improved efficiency, excessive time constraints and rating systems inherently carry the risk of compromising safety management and hygiene standards. In particular, high turnover and non-regular employment status obscure hygiene training and the allocation of responsibility, expanding food safety from an issue for individual companies to a challenge for the entire institutional design, including the platform.

Therefore, the OFD market must be viewed not only as a socio-economic infrastructure supporting urban employment and consumer convenience but also as an area where the labor structure inherent in gig work intersects with food safety risks, and thus requires institutional governance.

7.3 Food Safety Risks Specific to OFD

While China's OFD market has expanded rapidly, this growth has brought with it unique food safety risks that differ from those of the traditional food service industry. In this section, we examine the risks specific to OFD from four perspectives: (1) lack of visibility into food preparation areas, (2) unlicensed operations and "ghost" stores, (3) secondary contamination during delivery, and (4) the rating economy and information asymmetry.

7.3.1 The "Invisible Kitchen" Issue

The expansion of OFD has fundamentally transformed the way food and beverage services are provided. Its core characteristics are the "invisibility" of the cooking space and the weakening of oversight functions. In traditional dine-in settings, consumers could directly observe the restaurant's location, cleanliness, and the hygiene of staff, allowing a certain degree of social oversight to function naturally. However, with OFD, everything from ordering to payment and reviews is handled entirely through an app, and consumers cannot verify the actual cooking environment.

The "invisible kitchen" issue can be defined as a phenomenon in which the functions of monitoring, evaluation, and accountability regarding food safety are institutionally weakened because the cooking space is placed in a state where it cannot be directly observed by consumers or regulatory authorities. This phenomenon creates new tensions within the foundation of institutional trust formation, as discussed in this book.

Specifically, the following issues have been identified:

- Prepared food is left at room temperature for extended periods (including delivery delays)
- It is impossible to monitor in real time if kitchen hygiene deteriorates after initial inspection
- Oversight of the entire supply chain, including the delivery process, is fragmented

These issues have also been reported in existing research on online food delivery as a phenomenon in which food safety risks extend across the entire supply chain (Li et al., 2020).

Small-scale establishments and so-called "ghost kitchens," in particular, face little social pressure to maintain their exteriors or keep their interiors clean, as they either lack a physical storefront or do not expect walk-in customers. This structure has the potential to render traditional surveillance mechanisms, which rely on physical spaces, ineffective.

Furthermore, while OFD platforms (e.g., Meituan, Ele.me) require the online display of business licenses and store photos, this information is static and does not sufficiently guarantee daily hygiene standards. Regarding administrative oversight, while efforts to strengthen supervision of online food services are underway, led by the State Administration for Market Regulation, the sheer number of establishments and high turnover limit the comprehensiveness of on-site inspections (State Administration for Market Regulation, 2021).

Furthermore, the "invisible kitchen" issue is not merely a matter of insufficient information; it is exacerbated by its interaction with the review economy. Consumer reviews tend to focus primarily on taste and delivery speed, and hygiene management is rarely visible unless an incident occurs. As a result, institutional distortions may arise where incentives for hygiene management are not sufficiently reflected in market evaluations.

Thus, the "invisible kitchen" is a structural problem stemming from the spatial invisibility inherent to OFD. It weakens the social surveillance and face-to-face trust that implicitly functioned in traditional brick-and-mortar food service establishments, creating an environment that relies excessively on institutional trust. Therefore, ensuring food safety requires not merely the strengthening of hygiene standards, but a restructuring of institutional design, such as the visualization of cooking environments and the enhancement of information disclosure.

7.3.2 The Unlicensed Operations and "Ghost Kitchen" Issues

While the rapid expansion of the food delivery market has lowered barriers to entry, it has also brought to light issues such as unlicensed operations and so-called "ghost kitchens." Ghost kitchens refer to businesses that operate exclusively online without a physical storefront, or those whose business licenses or location information do not match their actual operations. These businesses can easily evade visible oversight in physical spaces and inherently pose food safety risks.

In China, obtaining and displaying a business license is mandatory for food operations; however, on OFD platforms, fraudulent practices such as reposting license images or misappropriating information from other establishments have become a problem. Regulations on online food and beverage services require third-party platforms to verify the food business permits of tenant food and beverage operators, register their names, addresses, and responsible persons, ensure the authenticity of permit information, and clarify food safety responsibilities (China Food and Drug Administration, 2017).

However, the rapid increase in the number of stores and high turnover rates have placed a significant burden on both pre-approval screening and post-approval supervision. In fact, on April 17, 2026, the State Administration for Market Regulation imposed administrative penalties totaling 3.597 billion yuan on seven major platforms: Pinduoduo, Meituan, JD.com, Ele.me (Taobao Flash Purchase), Douyin, Taobao, and Tmall. This was one of the largest penalties in the online food safety sector since the implementation of the Food Safety Law, with the primary reason for the crackdown cited as the failure to address illegal activities related to "ghost delivery" (Xinhuanet, 2026).

This case symbolically illustrates the current reality in which platforms' screening responsibilities and supervisory obligations are not functioning effectively. Furthermore, a platform's failure to properly fulfill its review obligations is not merely a matter of poor management; it is a serious issue that erodes the very foundations of food safety and market order. Specifically, a review system that relies solely on formal verification of certificates facilitates the entry of unlicensed operators and undermines basic food safety standards by overlooking substandard raw materials and inappropriate manufacturing environments. Furthermore, store listings containing false information or exaggerated claims undermine consumers' right to know and erode their trust, and may even lead to the improper handling of personal information. In addition, the multi-layered transaction structure facilitated by platforms fuels excessive price competition, leading to a normalization of declining quality under cost-cutting pressures, which in turn undermines the health of the industry as a whole.

On the platform side, while companies like Meituan and Ele.me have introduced AI screening and license verification systems, these are limited to formal checks of submitted documents, making it difficult to fully grasp actual cooking environments and operational realities. Online information is static and does not guarantee daily changes in hygiene conditions or personnel management systems.

Further complicating the issue is the ambiguity regarding where responsibility lies. In the event of a food safety incident, the direct party responsible is the restaurant that prepared the food; however, the contractual relationship with the consumer is established through the platform. This "dual structure" creates a separation between legal and social responsibility, raising new challenges from the perspective of consumer protection. Algorithmic governance in the platform economy, while increasing efficiency, has the potential to lead to a diffusion of responsibility and regulatory gaps (OECD, 2020).

The issue of unlicensed operations and "ghost stores" is not merely a collection of individual illegal acts, but a risk stemming from the structural characteristics of the online marketplace. Specifically, (1) low barriers to entry, (2) reliance on digital screening, (3) limited supervisory resources, and (4) the platform intermediary structure interact to institutionally create a lack of transparency regarding the true identity of operators. Therefore, resolving this issue requires not only stricter penalties but also institutional redesign, including real-time synchronization of business license information, enhanced data sharing, and clarification of joint liability for platforms. For the sound development of the OFD market, a clear legal framework addressing the fundamental question of "who is the responsible party" is indispensable, and this issue can be viewed as a symbolic case that compels the restructuring of food safety governance.

7.3.3 Secondary Contamination during Delivery

One of the food safety risks specific to OFD is secondary contamination caused by the "delivery process" that occurs between the time food is prepared and the time it is consumed. In traditional dine-in settings, preparation and consumption take place within the same space and within a short timeframe, so hygiene management was primarily focused on the kitchen itself. However, with OFD, prepared food is transported in an outdoor environment for a certain period of time, giving rise to new risk factors such as temperature control, container management, and human contact.

First is the issue of temperature control. From a food safety perspective, cooked food must be kept at an appropriate temperature, while refrigerated food must be maintained at a low temperature. However, delivery delays and long-distance transport in high-temperature environments increase the likelihood of bacterial growth.

Particularly during the summer or in times of traffic congestion, the risk of food being exposed to the danger zone for extended periods increases. Such temperature control challenges have also been identified in existing research on online food delivery as food safety risks associated with the extension of the supply chain (Li et al., 2020).

Second is the possibility of secondary contamination due to physical contact. Delivery personnel handle a large number of orders in a short period, repeatedly opening and closing bags and rearranging containers. If packaging is inadequate, cross-contamination may occur due to contact with the external environment or other foods. During the COVID-19 pandemic, measures such as contactless delivery and the use of tamper-evident seals were promoted; however, these were primarily introduced as infection control measures and have not been systematically institutionalized as part of food hygiene management. Regarding hygiene risks in the delivery process, multiple studies have pointed out that delivery equipment and insulated bags can serve as sources of contamination, emphasizing the importance of proper cleaning and management (Li et al., 2020).

Third, there is the relationship between algorithmic governance and working conditions. Major platforms such as Meituan and Ele.me strictly manage delivery times through real-time delivery management systems. The combination of piecework pays and time limits creates an incentive for delivery workers to prioritize speed above all else. Research on platform labor points out that while algorithmic management highly standardizes and accelerates the labor process, safety checks and work quality may be pushed to the background (Sun et al., 2021). In other words, the risks in the delivery process are not merely technical issues but are closely linked to the labor governance structures unique to the platform economy.

Furthermore, the allocation of responsibility in the event of an accident is complex. If food spoilage or contamination occurs during delivery, determining which party, the store, the delivery worker, or the platform, bears responsibility is difficult both legally and in practice. This ambiguity regarding liability may weaken incentives for proactive safety investments. A report by the International Labour Organization also points out that platform work is prone to the fragmentation of responsibility and institutional gaps (International Labour Organization, 2021).

Secondary contamination during the delivery process is a structural risk inherent in the OFD model, which involves extended timeframes and spatial movement. It requires a multi-layered response, including the development of temperature control technologies, standardized packaging specifications, hygiene training for delivery personnel, and a review of algorithm design. Food safety is no longer an issue confined solely to the kitchen; it has expanded into a governance issue encompassing the entire logistics network built by the platform.

7.3.4 *The Rating Economy and Information Asymmetry*

The rating economy refers to a system in which data such as ratings, reviews, and transaction histories provided by consumers and users are visualized and accumulated to reduce uncertainty and information asymmetry regarding the quality of goods and services, and in which this data influences the formation of trust and the allocation of resources (selection, competition, and rewards) in the market. In other words, it is a market coordination mechanism of the digital age in which, in addition to price and quality themselves, "rating information" mediates transactional decision-making and disciplines the behavior of firms and workers (Dellarocas, 2003; Resnick et al., 2006).

In the OFD market, this rating economy functions particularly strongly. Under conditions where consumers cannot inspect physical goods in advance, they select stores based on platform metrics such as star ratings, review comments, sales volume, and rankings. Since this evaluation information directly affects a store's visibility and order volume, it acts as an incentive for restaurants and delivery personnel to maintain quality and provide prompt service.

At the same time, this rating economy goes beyond merely providing information; when combined with algorithms, it functions as a "quasi-market governance mechanism" that influences exposure rankings and business opportunities. Therefore, it can be said that the formation of consumer trust in the OFD market is supported not by traditional face-to-face transactions or brand trust, but by "data-mediated trust" based on the review data accumulated on the platform (OECD, 2020).

However, these rating metrics do not necessarily accurately reflect the actual state of food safety. In this section, we examine the institutional risks specific to the OFD market from the perspectives of the review economy and information asymmetry.

First is the bias in the rating and review content. Consumer reviews tend to focus heavily on immediate, experiential factors such as taste, price, delivery speed, and packaging condition. On the other hand, specialized information, such as kitchen hygiene management and the safety of raw materials, is difficult to visualize unless an incident occurs. This has been pointed out in research on online food delivery as a potential issue where food safety information is not sufficiently reflected in consumer ratings (Li et al., 2020). Consequently, market ratings may not adequately signal hygiene standards.

Second, there is the issue of algorithmic ranking. Major platforms such as Meituan and Ele.me determine search rankings and recommended listings based on proprietary algorithms. However, the criteria for these calculations are not fully disclosed, and it has been pointed out that advertising spending and sales performance may influence them. In other words, information that consumers perceive as "popular" or "highly rated" is not necessarily a pure indicator of quality but may be influenced by commercial factors and platform strategies (OECD, 2020).

Third, there is the structural persistence of information asymmetry. In the OFD, while businesses have detailed knowledge of their own hygiene conditions and ingredient sourcing channels, consumers can only access the limited information displayed on the app. While the online posting of business licenses and photos provides a certain level of transparency, this remains static and formal information. It is difficult for outsiders to observe dynamic hygiene management conditions or daily compliance practices. As a result, self-regulation based on market mechanisms is limited.

Furthermore, the issues of rating manipulation and fake reviews cannot be ignored. In a highly competitive environment, behaviors such as the intentional purchase of high ratings or the suppression of low ratings may occur. Although the platform has implemented monitoring systems, it is difficult to eliminate these practices entirely. Such distortions in ratings weaken the structure that relies excessively on market-based trust rather than institutional trust.

While the rating economy serves as a central mechanism for building trust in the OFD market, its information structure is hardly designed to adequately reflect the specialized and latent risks associated with food safety. Therefore, rather than a simple star-rating system, a design that enhances institutional transparency, such as the visualization of hygiene inspection results, the disclosure of violation histories, and the integration of real-time oversight information, is required. Food safety issues on OFDs raise a fundamental question for digital market societies: to what extent can "trust through ratings" actually guarantee safety?

7.3.5 Food Safety Risks Specific to OFD and the Erosion of Institutional Public Goods

The risks specific to OFD, namely, (1) the lack of visibility into food preparation areas, (2) the issues of unlicensed operations and "ghost stores", (3) secondary contamination during delivery, and (4) the "rating economy" and information asymmetry, must be understood not merely as individual or technical challenges, but as structural risks that affect the very institutional foundation underpinning food safety. In other words, these four risks are interrelated and are eroding from within the foundation of trust that food safety, which should fundamentally be supplied as an "institutional public good", reliably provides.

Here, the term "institutional public good" refers not merely to public goods possessing non-excludability and non-rivalry (Samuelson, 1954), but also to the social foundations of trust and safety that are maintained and supplied by institutions. From the perspective of institutional economics, the smooth functioning of markets depends on institutional frameworks such as trust, norms, and rules (North, 1990), and food safety is likewise understood as a public value established under institutional conditions that transcend the efforts of individual companies.

First, the invisibility of the cooking space solidifies an information structure in which consumers cannot directly verify quality or hygiene, thereby eliminating the "visual supervision" and "social monitoring" that functioned in traditional face-to-face dining. Consequently, food safety becomes entirely dependent on ex post institutional guarantees, and if trust in the institutions declines, trust in the market as a whole is simultaneously undermined.

Second, the issues of unlicensed operations and "ghost stores" render the very conditions for accessing the system meaningless. While the business licensing system is intended to function as an "entry barrier" that allows only entities meeting safety standards to participate in the market, if this filter is effectively rendered ineffective due to superficial reviews and inadequate oversight on the platform, food safety, as an institutional public good, collapses at the supply stage.

Third, secondary contamination during the delivery process expands the scope of responsibility for food safety while simultaneously dispersing accountability. Even if food was appropriate at the preparation stage, if its quality deteriorates due to inadequate temperature control or hygiene management during transport, it becomes unclear whether responsibility lies with the store, the delivery person, or the platform. Such fragmentation of responsibility undermines the "consistency of responsibility" that is essential for maintaining institutional public goods (OECD, 2020).

Fourth, the review economy and information asymmetry cause the function of building trust, which should inherently be the responsibility of institutions, to be replaced by algorithms within the market; however, if those evaluation metrics are susceptible to manipulation, the very foundation of trust is distorted. Star ratings and reviews are primary sources of information for consumer decision-making; however, if false reviews and ranking manipulation become rampant, trust will collapse from within, ultimately leading to a decline in trust in the system (Li et al., 2020).

In summary, food safety issues in the OFD market must be addressed not as a matter of "managing individual risks," but rather in terms of "rebuilding trust as an institutional public good." In other words, food safety is not merely a matter of quality assurance, but a public value established through the interaction of institutions, markets, and technology; its erosion diminishes not only market efficiency but social welfare itself.

From this perspective, institutional redesign including the clarification of platform responsibilities, ensuring the authenticity of information, and enhancing oversight through data integration—is indispensable. The OFD market can thus be positioned as a prime example of an area that demands a reorganization of the supply mechanisms for institutional public goods.

7.4 Design of the OFD Food Safety Governance System

As clarified in Sect. 7.3, food safety risks in the OFD market cannot be attributed solely to the actions of individual actors; rather, they are structural problems arising from the interplay of platform structures, algorithmic governance, and information asymmetry. Therefore, addressing these risks requires more than just traditional store regulations or ex post supervision; it must be viewed as a restructuring of institutional design. In this section, we examine how to reconstruct food safety as an institutional public good from three perspectives: platform liability, algorithmic governance, and collaborative oversight.

7.4.1 Redefining Platform Responsibility

In China, the E-Commerce Law and food safety regulations impose obligations on platform companies to screen merchants, disclose information, and address illegal activities. As a result, platforms are positioned not merely as information intermediaries, but as entities bearing certain management responsibilities.

The essence of this institutional shift lies in the reorganization of "accountability." Specifically, because transactions in the OFD market are multi-layered, not only the merchants but also the platforms that design and operate the trading environment are considered entities that should bear structural responsibility for food safety. This indicates that platforms are increasingly taking on the character of providers of institutional public goods, in the sense that they function as governance mechanisms that determine market conditions (OECD, 2020).

However, the current system relies primarily on ex post liability and formal reviews, and has limitations in terms of dynamic hygiene management and understanding actual conditions. Going forward, a shift from mere "mandates" to "effective allocation of responsibility" is required.

7.4.2 Algorithmic Governance and Structural Responsibility

In the OFD market, factors such as store ranking, order allocation, and delivery time settings are determined by algorithms, which effectively function as governance mechanisms that define market structure and behavioral incentives.

While algorithms enhance efficiency, an excessive emphasis on speed or cost minimization may weaken incentives for ensuring safety and maintaining quality. For example, strict delivery time constraints can lead to neglect of temperature control and hygiene checks. Such issues should be understood as "structural risks" stemming from the designed incentive structure, rather than as deviations in individual behavior.

Therefore, food safety governance must extend beyond the detection of violations to include regulation and auditing of the algorithm design itself. In other words, institutional requirements such as algorithmic transparency, accountability, and the incorporation of safety metrics are essential.

7.4.3 Collaborative Oversight and Data-Driven Governance

Given the scale and complexity of the OFD market, there are structural limitations to supervision by government agencies alone. For this reason, China has recently placed greater emphasis on co-regulation by government agencies and platforms.

The core of this model lies in the division of roles and the utilization of data. The government assumes legal authority and ultimate responsibility, while platforms handle day-to-day monitoring and data provision. Transaction data, user reviews, and complaint information accumulated by platforms serve as critical resources for identifying high-risk merchants. This enables risk-based supervision, allowing for the targeted allocation of regulatory resources.

However, co-regulation is not merely a means of improving efficiency; it is also a redesign of the mechanism for supplying institutional public goods. In other words, the key lies in how to combine the state's legal enforcement power with the platforms' technological governance capabilities while ensuring transparency and accountability. To achieve this, it is essential to clarify data-sharing rules, introduce third-party audit systems, and strengthen the disclosure of violation information.

7.4.4 Rebuilding Food Safety as an Institutional Public Good

Taking all of the above into account, food safety governance in the OFD market should be understood not merely as a tightening of regulations, but as a "redesign of institutional public goods." In other words, food safety cannot be maintained through the voluntary actions of market participants alone; it can only be sustainably ensured through governance that integrates institutions, technology, and the market.

From this perspective, the direction of institutional design can be summarized in the following three points.

First is clarifying platform liability and institutionalizing joint liability. Second is ensuring transparency and accountability in algorithmic governance. Third is the establishment of data-driven oversight through collaboration between government and platforms. The OFD market is at the forefront of trust provision in the digital economy, and food safety issues serve as a litmus test for the success of its institutional design. Therefore, the restructuring of governance in this market transcends the scope of mere industrial policy and can be positioned as a prime example illustrating the evolution of mechanisms for supplying institutional public goods.

7.5 Policy Implications: Reconstructing the Three-Layer Trust Model in the Age of Invisible Risks

While the digital economy reduces transaction costs and improves efficiency, it also possesses a structural characteristic that embeds risks within algorithms and data processing, reconfiguring them as "invisible risks." Food safety issues in OFD markets demonstrate that these invisible risks are transforming the very mechanisms of trust formation.

Based on the analysis in this chapter, the trust structure underpinning food safety must be reformulated from the perspective of this book's three-layer trust model, psychological trust, institutional trust, and structural trust.

First is the weakening of psychological trust. In traditional face-to-face dining, consumers formed sensory and experiential trust by directly observing the store environment and the hygiene of employees. However, in OFD, the cooking space is rendered invisible, and trust shifts to "indirect trust" that relies on reviews and ratings. As a result, trust increasingly depends on subjective and manipulable information, which can lead to a disconnect from the actual state of food safety.

Second, there is an increased reliance on institutional trust. In an invisible environment, consumers have no choice but to place their trust in government regulations and platform vetting systems. However, as evidenced by "ghost stores" and the limitations of oversight, when the effectiveness of these systems wavers, the very foundation of trust becomes unstable. In other words, while institutional trust becomes a central element that must be strengthened, its vulnerability is simultaneously exposed.

Third is the centralization of structural trust. In OFD markets, search rankings, recommendation algorithms, and delivery time settings dictate the terms of transactions, meaning that trust depends not on individual actors but on platform design. This signifies a shift in the object of trust from "people" to "systems." Algorithms are not merely technology; they function as governance mechanisms that determine resource allocation.

To summarize, food safety in the digital age is characterized by a reconfigured balance among the three layers of trust. Specifically, psychological trust has relatively diminished, reliance on institutional trust has increased, and structural trust now plays a decisive role. This shift indicates that the foundation of trust is transitioning from "face-to-face observation" to "institutions and design."

Consequently, the policy implications extend beyond the management of individual risks to the redesign of these three layers of trust. In particular, ensuring the transparency and explainability of algorithms, clarifying platform responsibility, and integrating institutional oversight with data governance are essential. At the same time, it is also important to supplement psychological trust through measures such as the visualization of hygiene information.

The OFD market is a microcosm of modern society, where trust is shifting from intuition to data, institutions, and structures, and its institutional design points the way for trust governance across the entire digital economy.

References

Alibaba Group. (2024). *Fiscal Year 2023 Annual Report*. https://static.alibabagroup.com/reports/fy2023/ar/ebook/en/index.html

Calo, R., & Rosenblat, A. (2017). The taking economy: Uber, information, and power. *Columbia Law Review, 117*(6), 1623–1690. https://columbialawreview.org/content/the-taking-economy-uber-information-and-power/

China Food and Drug Administration. (2017). *Wangluo canyin fuwu shipin Anquan jiandu guanli banfa* [Measures for the supervision and administration of food safety in online catering services]. https://sjfg.samr.gov.cn/law/file//pdf/3235243/1663580053364.pdf (In Chinese).

China Internet Network Information Center. (2023). *The 51st Statistical Report on China's Internet Development*. China Internet Network Information Centre. (In Chinese).

Dellarocas, C. (2003). The digitization of word of mouth: Promise and challenges of online feedback mechanisms. *Management Science, 49*(10), 1407–1424. https://doi.org/10.1287/mnsc.49.10.1407.17308

International Labour Organization. (2021). *World employment and social outlook 2021: The role of digital labour platforms in transforming the world of work*. https://www.ilo.org/publications/flagship-reports/role-digital-labour-platforms-transforming-world-work

Jia, P. (2020). Obesogenic environment and childhood obesity. *Obesity Reviews, 22*(S1), e13158. https://doi.org/10.1111/obr.13158

Kasperson, R. E., Renn, O., Slovic, P., Brown, H. S., Emel, J., Goble, R., Kasperson, J. X., & Ratick, S. (1988). The social amplification of risk: A conceptual framework. *Risk Analysis, 8*(2), 177–187. https://doi.org/10.1111/j.1539-6924.1988.tb01168.x

Li, C., Mirosa, M., & Bremer, P. (2020). Review of online food delivery platforms and their impacts on food safety. *Trends in Food Science & Technology, 96,* 200–210. https://doi.org/10.3390/su12145528

Meituan. (2024). *Annual report 2023*. https://media-meituan.todayir.com/20240429080801261185051_en.pdf

National Bureau of Statistics of China. (2024). *China statistical yearbook 2024*. China Statistics Press.

North, D. C. (1990). *Institutions, institutional change and economic performance*. Cambridge University Press. https://doi.org/10.1017/CBO9780511808678

OECD. (2020). *Competition in digital platform markets*. https://www.oecd.org/daf/competition/competition-in-digital-platform-markets-2020.pdf

Resnick, P., Zeckhauser, R., Swanson, J., & Lockwood, K. (2006). The value of reputation on eBay: A controlled experiment. *Experimental Economics, 9*(2), 79–101. https://doi.org/10.1007/s10683-006-4309-2

Samuelson, P. A. (1954). The pure theory of public expenditure. *The Review of Economics and Statistics, 36*(4), 387–389. https://doi.org/10.2307/1925895

State Administration for Market Regulation. (2021). *Measures for the supervision and administration of food safety in online catering services.*

Sun, P., Chen, J. Y., & Rani, U. (2021). From Flexible Labour to 'Sticky Labour': A tracking study of workers in the Food-Delivery Platform Economy of China. *Work Employment and Society, 37*(2), 412–431. https://doi.org/10.1177/09500170211021570

U.S. Department of Justice. (2010). *Horizontal merger guidelines*. https://www.justice.gov/atr/horizontal-merger-guidelines-08192010

Xinhuanet. (2026, April 17). *Pingduoduo, Meituan, Jingdong, Eleme, Douyin, Taobao, Tianmao 7jia dianshang pingtai bei famei 35.97 yiyuan* [Seven e-commerce platforms, Pinduoduo, Meituan, JD.com, Ele.me, Douyin, Taobao, and Tmall, have been fined a total of 3.597 billion yuan] https://www.news.cn/20260417/b9a3d4983bae4abebcf9f5df2e9e12c6/c.html (In Chinese).

Chapter 8
The Institutionalization of Food Safety as a Public Good and a Sustainable Society: Integration of the Three-Layer Trust Model and Policy Recommendations

Weisheng Zhou

8.1 Introduction

The objective of this chapter is to integrate the institutional transformation of food safety governance discussed in Chap. 6 with the structural risks associated with the expansion of the OFD market analyzed in Chap. 7, and to propose policy directions for rebuilding trust from the perspective of building a sustainable society in the digital economy.

Food safety issues are not merely a matter of compliance with hygiene standards or enforcement against violations; they are socio-structural challenges in which information, institutions, technology, and markets interact. Because consumers cannot fully verify the safety or quality of food in advance, food inherently involves information asymmetry. As Akerlof (1970) demonstrated, information asymmetry leads to adverse selection in the market, which can result in the influx of low-quality goods and a decline in overall market trust. Furthermore, food is a classic trust good, and consumers are often unable to fully verify its quality even after purchase (Darby & Karni, 1973).

While the advancement of the digital platform economy has the potential to mitigate this information asymmetry, it also creates new invisible risks. Reviews, ratings, rankings, and recommendation algorithms assist consumers in their decision-making; however, if the criteria for their display or the processing procedures remain opaque, the reliability of the information actually diminishes. As Pasquale (2015) argues, in markets organized by algorithms, accountability for the information processing procedures, not just the disclosure of information, is crucial.

W. Zhou (✉)
College of Policy Science, Ritsumeikan University, Kyoto, Japan
e-mail: zhou@sps.ritsumei.ac.jp

W. Zhou (ed.), *Food Safety in China*, https://doi.org/10.1007/978-981-92-3741-8_8

Furthermore, food platforms, including OFD markets, possess the characteristics of two-sided markets. As Rochet and Tirole (2003) demonstrated, platforms act as intermediaries between multiple user groups and are entities that simultaneously coordinate prices, information, and transaction conditions. From the perspective of food safety, platforms are not merely intermediaries; they function as quasi-public governance entities that shape market order through store vetting, ranking of listings, review management, complaint handling, and responses to violations.

Therefore, the first challenge is the redesign of institutional trust. It is necessary not only to strengthen administrative oversight but also to clarify the responsibilities of platform companies and establish a mechanism in which the government, businesses, platforms, and consumers jointly support food safety. The second challenge is the enhancement of digital transparency. It is necessary to make violation histories, hygiene permits, traceability information, and evaluation criteria visible, thereby creating an information environment that allows consumers to make rational judgments. The third challenge is integration with sustainability. Food safety is linked not only to hygiene management but also to environmental impact, packaging waste, delivery labor, and relationships with local communities.

Based on the above, this chapter redefines food safety, shifting its focus from a framework of "accident prevention" to the "reconstruction of social trust capital." Trust is not merely a psychological emotion but a social foundation supported by the integrated design of institutions, technology, information, and norms. The purpose of this chapter is to present a new model of equilibrium among the state, the market, and society in the digital economy as a policy recommendation.

8.2 Current Status and Challenges in Global Food Safety (Quality Safety)

8.2.1 Current Status

Global food safety is becoming increasingly important, extending beyond the mere assurance of food supply to encompass "quality safety", that is, protecting consumers from microbial contamination, chemical substances, pesticide residues, allergens, and mislabeling. According to estimates by the World Health Organization, contaminated food causes approximately 600 million cases of illness and about 420,000 deaths worldwide each year, indicating that food safety has a serious impact on public health and socioeconomic conditions (World Health Organization, 2015, 2024). Furthermore, economic losses due to foodborne diseases in low- and middle-income countries are estimated to reach approximately $110 billion annually, positioning food safety not only as a health issue but also as a development challenge (Jaffee et al., 2019).

Table 8.1 presents key indicators on global food safety based on WHO data. This table demonstrates that the current state of food safety worldwide remains serious. The fact that approximately 600 million people annually, or about 10% of the global population, suffer from foodborne illnesses, resulting in approximately 420,000 deaths, indicates that food safety is not a problem confined to specific regions but rather a structural risk with global reach. Of particular note is that children under five account for about 30% of these deaths, demonstrating that food safety issues disproportionately affect socially vulnerable groups.

Furthermore, the disease burden of approximately 33 million DALYs suggests that food safety is not merely a temporary health issue but one that entails long-term health losses and socioeconomic costs. Moreover, given that foodborne diseases encompass more than 200 types, it is clear that the risk structure is multifaceted and cannot be resolved by a single regulation or technology.

In light of the above, it can be understood that the essence of food safety issues lies not merely in a lack of hygiene management, but in a "lack of institutional trust" stemming from information asymmetry and imperfections in oversight systems. In other words, food safety cannot be fully ensured by market mechanisms alone, and the core challenge lies in how to provide trust as an institutional public good. In this regard, food safety is positioned as an integrated governance issue closely linked to multiple SDGs, particularly Goal 2 (Zero Hunger), Goal 3 (Good Health and Well-being), Goal 12 (Responsible Consumption and Production), and Goal 16 (Peace, Justice, and Strong Institutions).

Therefore, this table highlights the need to re-examine food safety issues not merely as a collection of individual risks, but as a matter of building trust through the integration of institutions, markets, and technology.

Table 8.1 Key indicators of global food safety issues

Indicator	Global statistics	Interpretation (qualitative safety perspective)
Number of foodborne Illness Cases	Approx. 600 million/year (about 1 in 10 people)	Food safety is a globally persistent risk
Number of deaths	Approx. 420,000 deaths/year	Major public health issue
Proportion of deaths among children under 5 years old	Approx. 30% (about 125,000 people)	Concentrated risk for vulnerable populations
Disease burden (DALYs)	Approx. 33 million DALYs/year	Large long-term health and social costs
Number of major diseases	More than 200 types	Diverse risk structures (biological, chemical)
Major disease composition	Diarrheal diseases account for the majority	Directly linked to basic sanitation and institutional problems
Economic loss (estimated)	Approx. 110 billion USD/year (low- and middle-income countries)	Significant impact on development and growth

Source compiled from World Health Organization (2015, 2024)

8.2.2 The Structure of Global Food Safety Risks

The structure and challenges of global food safety (quality safety) can be summarized as follows:

First, food safety risks are becoming increasingly complex. Modern food supply chains are highly specialized and operate across national borders, creating a structure in which regulatory failures in a single country can lead to international risks. Diverse risks exist at every stage, from production and processing to distribution and sales, and these must be managed in an integrated manner. The Codex Alimentarius Commission emphasizes preventive management based on Good Hygiene Practices (GHP) and Hazard Analysis and Critical Control Points (HACCP) as fundamental principles of food safety management, highlighting the necessity of risk management across the entire food chain (Codex Alimentarius Commission, 2022).

Second, disparities in institutional capacity among nations remain a major challenge. While developed countries have well-established inspection systems, risk analysis, and traceability systems, many low- and middle-income countries lack sufficient inspection capacity, cold chain infrastructure, and enforcement of legal frameworks. Consequently, there are significant differences in response capabilities among countries, even when facing the same food safety risks. Jaffee et al. (2019) point out that improving food safety requires a comprehensive approach including institutional capacity building, private sector engagement, and consumer education, rather than merely strengthening regulations.

Third, there are issues of information asymmetry and trust. Food is a "credence good" in which consumers cannot fully assess safety or quality before purchase, resulting in an information gap between sellers and buyers. In this regard, Akerlof's (1970) "lemons market" theory demonstrates that information asymmetry can lead to a deterioration in market quality. In the food market as well, when information regarding quality is insufficient, there is a risk that low-quality products will enter the market or that mislabeling will occur, leading to a decline in trust in the market as a whole. For this reason, ensuring transparency of information through labeling systems, certification systems, traceability, and third-party audits is essential for food safety.

Fourth, the emergence of new risks associated with digitalization. The expansion of e-commerce (EC) and food delivery services has accelerated and diversified food distribution. While this improves consumer convenience, it also gives rise to new challenges, such as the difficulty in identifying sellers, uncertainty regarding temperature control, and ambiguity regarding liability. In this environment, management relying solely on traditional ex post supervision is insufficient; there is a need for data-driven real-time monitoring and advanced traceability (Food and Agriculture Organization of the United Nations, 2023).

Fifth, there is a growing need to provide food safety as an "institutional public good." The WHO Food Safety Strategy (2022–2030) aims to establish a preventive and integrated food safety system based on scientific evidence. Furthermore, the FAO positions food safety as a core element of sustainable food systems and places

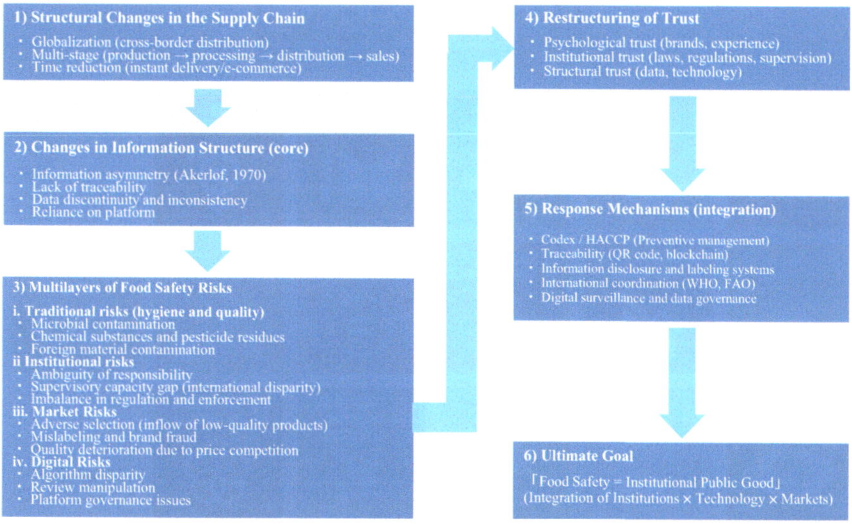

Fig. 8.1 Structural model of global food safety risks (qualitative safety) created by the author

emphasis on international cooperation and strengthening institutional capacity (Food and Agriculture Organization of the United Nations, 2023).

Figure 8.1 presents food safety risks as a dynamic process: "structural changes in supply chain → changes in information structure → multilayers of food safety risks → restructuring of trust →ultimate goal: institutional public good". In particular, with the advancement of e-commerce and digitalization, global food safety, especially issues related to quality safety, has evolved beyond mere hygiene management or the concerns of individual companies. It has become a complex governance challenge encompassing the structure of the entire supply chain, institutional capacity, information transparency, and the utilization of digital technology. Going forward, it will be necessary to establish food safety as an institutional public good through the integration of international standards based on Codex, preventive management based on HACCP, risk analysis, information disclosure, and traceability enabled by digital technology.

8.3 Reintegrating Food Safety and the UN Sustainable Development Goals

Food safety has traditionally been discussed as part of domestic public health and consumer protection policies. However, from the perspective of sustainable development, it is not a standalone policy area but rather a cross-cutting issue closely linked to food supply systems, health security, and resource circulation.

Fig. 8.2 United Nation Sustainable Development Goals (SDGs)

The United Nations Sustainable Development Goals (SDGs), as shown in Fig. 8.2, are international common goals adopted by the United Nations in 2015 and are formally a framework based on "Transforming our world: The 2030 Agenda for Sustainable Development" (United Nations, 2015). Their purpose is to realize a sustainable and inclusive society by 2030, and they are considered goals that all countries regardless of whether they are developed or developing, should strive to achieve. Within this framework, food, health, and production and consumption are positioned in an integrated manner, necessitating a reinterpretation of food safety as a central component of the SDGs (United Nations, 2015). In particular, the expansion of China's OFD market is leading to a restructuring of urban food supply systems while simultaneously bringing to light new issues related to sanitation, waste, and resource utilization. In this section, we examine the reintegration of food safety from the perspectives of SDGs 2, 3, 12, and 16.

8.3.1 SDG 2 (Zero Hunger): Ensuring Food Security

SDG 2 aims to "end hunger, achieve food security and improved nutrition, and promote sustainable agriculture" (United Nations, 2015). The term "food security" used here does not simply refer to quantitative supply, but is also a concept that includes sustained access to safe and nutritious food (Food and Agriculture Organization of the United Nations, 2023).

A lack of food safety undermines effective food security through health risks, even when access to food is available. According to the World Health Organization

(2015), approximately 600 million people worldwide contract foodborne illnesses each year, with the burden concentrated in low- and middle-income countries. Therefore, ensuring food safety is a prerequisite for reducing hunger and improving nutrition.

While China's OFD industry has expanded food access in urban areas, it also has the potential to widen disparities in hygiene management through small-scale operators and the "invisible kitchen" issue. In this sense, achieving SDG 2 requires the strengthening of unified hygiene standards and traceability systems in parallel with the expansion of digital distribution.

8.3.2 SDG 3 (Good Health and Well-being): Reducing Food Safety Risks

SDG 3 aims to "ensure healthy lives and promote well-being for all at all ages" (United Nations, 2015). Food safety is the foundation for the prevention of infectious and chronic diseases and for improving nutrition. Foodborne diseases remain a global public health challenge, and the risks are becoming increasingly complex, particularly in societies undergoing rapid urbanization and the externalization of dietary habits (World Health Organization, 2015).

Furthermore, the expansion of the OFD economy may accelerate the externalization of dietary habits and create new health challenges, including issues related to nutritional balance, food choices, and working conditions for delivery workers. Therefore, food safety policies must go beyond merely preventing microbial contamination and be integrated with comprehensive health policies, such as ensuring transparency in nutritional labeling, promoting health-conscious menus, and safeguarding worker safety.

In addition, working conditions in the platform economy are also linked to SDG 3. Overwork among gig workers and the risk of traffic accidents are factors that threaten workers' health and impact the sustainability of the entire food supply system. Health should be redefined as a concept that encompasses not only consumers but also occupational safety on the supplier side.

8.3.3 SDG 12 (Sustainable Consumption and Production): Transforming Food Systems

SDG 12 aims to "ensure sustainable consumption and production patterns" (United Nations, 2015). In the food sector, reducing food loss, improving resource efficiency, and enhancing waste management are key challenges.

The rapid expansion of the OFD market has led to an increase in single-use containers and higher energy consumption during delivery, in exchange for greater convenience. United Nations Environment Programme (2021) points out that to achieve sustainable consumption, the introduction of Extended Producer Responsibility (EPR) and circular economy models is essential.

Sustainable production and consumption are also crucial from a food safety perspective. Excessive cost competition can suppress investment in hygiene management and potentially lead to a decline in safety standards. Conversely, transparent supply chains and responsible production and consumption simultaneously promote quality improvement and build trust.

8.3.4 SDG 16 (Peace, Justice and Strong Institutions): Strengthening Regulation and Governance

Goal 16 aims to establish fair and reliable systems, improve institutional quality, and build a foundation of trust in society. Food safety should be understood not merely as a collection of hygiene management practices, but as a public good that depends on institutional quality and governance capacity. In other words, food safety is determined not by whether regulations exist, but by how consistently, transparently, and accountably those regulations are implemented. What is crucial here is the provision of trust through institutions; within the framework of this book, this corresponds to trust as an institutional public good. First is the clarification and consistency of the legal system. Food safety regulations are prone to undermining trust due to unclear standards, duplication, or excessive discretion. Therefore, it is essential to construct a unified and hierarchically organized legal system and minimize variations in application between central and local governments. This enhances the predictability of regulations and strengthens institutional trust among both businesses and consumers.

Second is the strengthening of enforcement capacity. No matter how sophisticated the system may be, trust cannot be established if enforcement is inconsistent. The development of inspection systems, the introduction of risk-based supervision, and reliable sanctions for violations form the foundation of institutional trust. At the same time, internal controls and audit mechanisms to curb selective or arbitrary enforcement are also indispensable. Third is the institutionalization of transparency and accountability. Disclosing food safety information (inspection results, violation cases, recalls, etc.) and making decision-making processes transparent directly contributes to the formation of psychological trust. In particular, with the advancement of digitalization, traceability and online disclosure should be incorporated as institutional obligations. Regarding supervision using algorithms, it is also necessary to ensure a certain level of explainability and auditability. Fourth is the establishment of co-governance. In modern food distribution, particularly in the OFD market, platform companies function as de facto regulatory bodies. Therefore, it is necessary to institutionalize a multi-stakeholder governance system in which government, businesses, and consumers share roles, and to clarify the allocation of responsibility and coordination mechanisms. This enhances the effectiveness of institutional trust by strengthening structural trust.

In summary, food safety serves as a cross-cutting foundation linking Goals 2, 3, 12, and 16, and requires policy design that integrates quantitative supply, health security, and resource circulation. In a sustainable society, food safety should be positioned not merely as "accident prevention" but as a core element of "institutionalized sustainability."

8.4 Policy Challenges in Rebuilding Trust in Food Safety in China

8.4.1 Enhancement in Information Disclosure

Enhancing information disclosure is a core policy challenge in rebuilding trust in food safety in China. Food safety is a classic trust good; consumers cannot fully verify quality either at the time of purchase or after consumption (Darby & Karni, 1973). Consequently, information asymmetry tends to lead to market failure, making institutionally guaranteed transparency indispensable (Akerlof, 1970).

In China, efforts have been made to publish cases of violations, disclose sampling inspection results, and release corporate credit information through the development of food safety legislation. However, with the expansion of the digital platform economy, the traditional model centered on public announcements is becoming less effective in building sufficient trust. Particularly in the OFD market, information such as store registration details, hygiene permits, past violation histories, and consumer complaint data is scattered across administrative departments and platforms, and this fragmentation of information is undermining transparency.

The first challenge is improving real-time data availability and data integration. It is necessary to institutionalize accountability not only for the disclosure of data itself but also for the criteria used in violation risk assessments, the logic behind ranking displays, and the methods used to calculate store ratings. As Pasquale (2015) points out, in markets mediated by algorithms, it is not only the volume of information disclosed that matters, but also how that information is selected and presented.

The second challenge is ensuring the publication of machine-readable data and interoperability. Integrating administrative inspection results, corporate credit information, and consumer complaint data and publishing them in a format that third-party organizations and researchers can analyze will strengthen external oversight functions. This signifies a shift from government-only supervision to social co-supervision.

The third challenge is the obligation of platform companies to engage in proactive disclosure. In two-sided markets, platforms act as gatekeepers of information flow and shape consumer behavior through rating systems and recommendation algorithms (Rochet & Tirole, 2003). Therefore, it is necessary to institutionalize structural transparency by disclosing the methods for calculating hygiene evaluation scores, the criteria for handling non-compliant establishments, and the complaint response processes.

Overall, the advancement of information disclosure represents a shift from "quantitative expansion" to "qualitative transformation." Evolving from fragmented disclosures to integrated, accountable disclosure, and from static announcements to dynamic data sharing, is essential for rebuilding trust in food safety in China.

8.4.2 Establishing a Risk Communication System as a Standard Practice

To rebuild trust in the field of food safety, it is essential not only to enhance the disclosure of information but also to institutionalize risk communication as a routine practice. Risk communication refers to the process of two-way information sharing among the government, businesses, experts, and the public regarding the nature of risks, uncertainties, and response strategies. The World Health Organization & Food and Agriculture Organization of the United Nations (2016) define food safety risk communication as a practical principle that underpins risk management.

Food safety is a field fraught with scientific uncertainty, and there is no such thing as zero risk. Therefore, it is not sufficient for the government to merely present scientific assessments; an institutional mechanism is needed to explain these assessments in a way that consumers can understand and accept. Frewer (2004) demonstrates that effective risk communication is not merely the transmission of information, but a process involving trust-building and policy acceptability.

The first policy challenge is the establishment of institutionalized dialogue mechanisms in peacetime. Rather than relying on one-sided explanations after an incident occurs, it is necessary to institutionalize regular public hearings, expert meetings, and online briefings, and to shift toward a multi-stakeholder participatory model that includes consumer groups and the media.

The second challenge is the development of an immediate response system in the digital environment. In the OFD market, information spreads instantaneously through social media and platform reviews. If misinformation or unverified information spreads, the absence of a swift and transparent official explanation could lead to growing distrust and panic. Therefore, it is crucial for government agencies and platform companies to collaborate to establish a unified information dissemination channel and a fact-checking system.

The third challenge is bridging the gap between expert knowledge and public understanding. While scientific assessments of food safety are highly specialized, trust cannot be established unless the results are understood by the public. System design that prioritizes comprehensibility is required, including visual materials, concise risk disclosures, and explanations of test results via QR codes.

Overall, to rebuild trust in China, it is necessary to shift from crisis-response public relations to an institutionalized model of routine risk dialogue. This will connect scientific rationality with social acceptability, serving as the foundation for sustainably supporting institutional trust in the digital age.

8.4.3 *Education and Improving Consumer Literacy*

When it comes to rebuilding trust in food safety in China, improving institutional frameworks and enhancing information disclosure alone are not sufficient. Ultimately, it is consumers who interpret risk information and make choices about their actions; therefore, improving consumer literacy is a prerequisite for ensuring that institutional trust is effective. Particularly in the digital platform economy, where reviews, rating scores, rankings, advertisements, and promotional information are intricately intertwined, the ability to interpret this information is directly linked to consumer safety.

Food safety is a classic trust good, and consumers cannot fully verify quality either before or after purchase (Darby & Karni, 1973). Consequently, they rely heavily on institutional guarantees and third-party certification. However, as Akerlof (1970) pointed out, in markets with significant information asymmetry, if consumers cannot sufficiently understand and evaluate information, it may lead to a decline in overall market quality.

The first challenge is the systematization of basic education on food safety. To prevent foodborne illnesses, proper storage, reheating, and hygiene practices at the consumer level are essential. In urban societies where the use of online food delivery services has become the norm, knowledge of consumption behaviors that takes into account delivery times, storage conditions, and reheating methods is crucial. It is necessary to disseminate basic knowledge of food safety through school education and community awareness campaigns.

The second challenge is strengthening digital information literacy. As Pasquale (2015) argues, algorithmic rankings and recommendations are not neutral information but rather the results of selections based on specific computational logic. Unless consumers can understand the meaning of rating scores, the reliability of reviews, and the structure of advertising displays, formal transparency will not translate into substantive trust.

The third challenge is the promotion of participatory oversight. As Frewer (2004) demonstrates, citizen participation in risk communication strengthens trust-building. In the field of food safety as well, establishing systems that allow consumers to participate in oversight processes through reporting, evaluation, and feedback will lead to the realization of social co-governance.

Overall, education and the improvement of consumer literacy serve as the foundation for transforming institutional trust into social practice. For the sustainable improvement of food safety in China, a long-term educational strategy to enhance citizens' judgment capabilities is indispensable, in addition to legal regulations and technological innovation.

8.4.4 Food Industry Strategy and Brand Trust Building

8.4.4.1 Quality Assurance and Differentiation Strategies

In the food market, quality assurance is not merely a matter of regulatory compliance; it is a strategic resource that enables companies to build long-term brand trust. Food is a classic trust good, and consumers cannot fully verify its quality either at the time of purchase or after consumption (Darby & Karni, 1973). This structural characteristic makes the market prone to failure due to information asymmetry, implying that the visibility of quality assurance becomes a source of competitive advantage for firms (Akerlof, 1970).

First, quality assurance serves as a mechanism for building reputational capital. Klein and Leffler (1981) argued that firms that consistently deliver high quality have an incentive to maintain that quality by securing future profits. While food companies' investments in implementing HACCP, obtaining third-party certification, and strengthening internal audit systems entail short-term cost increases, they enhance brand value and mitigate the risk of adverse selection in the long term.

Second, transparency is crucial as a differentiation strategy. In the digital economy, ratings and reviews on platforms directly influence consumer behavior. When companies voluntarily disclose information on raw materials, origin traceability, and inspection results, this mitigates information asymmetry and leads to differentiation from competitors.

Third, it is a strategy to integrate quality and brand. A brand is not merely a name but a cumulative representation of quality expectations. Once a reputation is established, consumers become willing to pay a price premium. Consequently, investment in quality achieves competitive differentiation and serves to prevent a decline in safety standards resulting from price competition (Klein & Leffler, 1981).

Overall, quality assurance is not a cost but a strategic investment and a core element of differentiation. In highly competitive environments such as China's OFD market, the shift from price-driven to quality-driven competition is the key to simultaneously improving food safety standards and building brand trust.

8.4.4.2 Restoring Trust in Domestic Brands

Food safety issues in China have gone beyond mere scandals involving individual companies, leading to a decline in trust in domestic brands as a whole. Past major food safety incidents have intensified consumer distrust of domestic products, leading to a preference for foreign brands and an expansion of cross-border shopping. This phenomenon demonstrates that reputation acts collectively in markets characterized by significant information asymmetry. Misconduct by a few companies generates externalities that damage the reputational capital of domestic brands as a whole (Akerlof, 1970).

Since food is a trust good and quality verification is difficult, consumers tend to use brand and country-of-origin images as proxy indicators of quality (Darby & Karni, 1973). Therefore, restoring trust in domestic brands requires not only the efforts of individual companies but also a comprehensive response that includes institutional guarantees and a national brand strategy.

First, the integration of institutional guarantees and branding is essential. To restore trust in domestic brands as a whole, institutional designs are required that clearly highlight reputable companies while imposing strict sanctions on non-compliant ones. A unified certification system, a credit rating system, and the publication of blacklists contribute to the reconstruction of the reputation market.

Second, the institutionalization of transparency and the enhancement of data disclosure are crucial. By institutionalizing raw material traceability, the disclosure of inspection results, and immediate notification of recall information, brands will be linked not merely to an advertising image but to verifiable quality assurance.

Third, a national-level quality strategy and industrial upgrading are essential. Restoring trust is not merely a crisis response; it must be linked to the upgrading of the industrial structure. Adding high value-added features, promoting standardization, and aligning with international standards will simultaneously enhance the competitiveness of domestic brands and restore trust.

In summary, restoring trust in domestic brands is achieved through quality improvements by individual companies, the strengthening of institutional guarantees, increased transparency, and the integration of national quality strategies. This is not merely a matter of improving image; it is a process of rebuilding reputational capital through the interplay of institutions and the market, and it is an indispensable task for the sustainable development of China's food industry.

8.5 The Feasibility of Establishing a "China Food Safety Model"

Food safety governance in China has undergone a rapid institutional transformation, triggered by a series of major incidents, through the strengthening of legal frameworks, the reorganization of supervisory systems, and the introduction of digital technologies. This series of changes is not merely an extension of institutional imitation or regulatory tightening; rather, it suggests the possibility of constructing a unique governance system tailored to China's political and economic structure and the characteristics of its digital society, namely, a "Chinese food safety model." In this section, we reformulate this model based on the three-layered trust model, comprising psychological, institutional, and structural dimensions, presented in this book, as shown in Fig. 8.3.

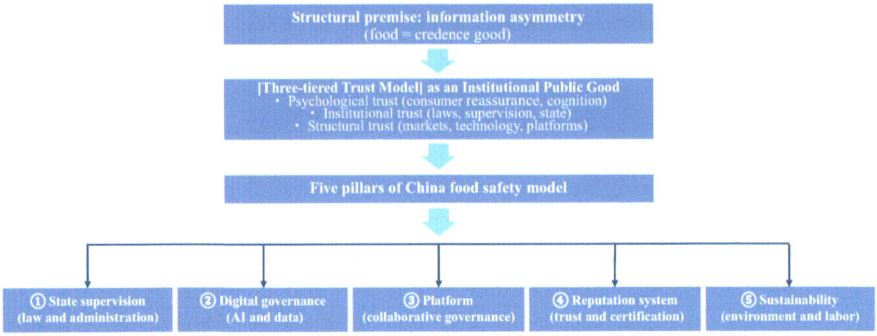

Fig. 8.3 China food safety model created by the author

First, the theoretical starting point of the Chinese model is the recognition that food markets inherently involve structural information asymmetry. As Akerlof (1970) demonstrated, markets where quality cannot be identified ex ante are prone to adverse selection, and as Darby and Karni (1973) pointed out, food is a classic trust good. For this reason, food safety cannot be adequately ensured by market mechanisms alone, and institutional intervention is indispensable. Based on this structural premise, food safety must be redefined not merely as a quality issue, but as a "problem of trust provision."

Second, a defining feature of China lies in the combination of centralized administrative capacity and digital governance capabilities. Through the introduction of big data oversight, credit information systems, and online monitoring platforms, the country is transitioning from traditional, post-hoc enforcement-centered supervision to preventive, real-time oversight. These elements function as core components of structural trust within the three-tier model presented in this book, supporting the institutionalization of trust through information visualization and traceability. However, as Pasquale (2015) points out, the advancement of algorithmic governance simultaneously entails new risks, such as the creation of black boxes and a lack of accountability.

Third, the Chinese model implies a shift from administration-led to collaborative governance. As demonstrated by Rochet and Tirole's (2003) two-sided market theory, platforms are not merely intermediaries but institutional actors that shape market order. Particularly in the OFD market, platform companies perform quasi-regulatory functions through store vetting, hygiene standard management, and the operation of rating systems. This role of the platform is a crucial component of structural trust, and when combined with institutional trust provided by the government, it forms a higher-order governance system. Therefore, the key to the Chinese model lies in institutionalizing coordinated oversight between the government and platforms.

Fourth, it involves the integration of reputational capital and institutional trust. Klein and Leffler (1981) demonstrated that a firm's incentive to maintain quality depends on the risk of losing future profits. In China, this reputation mechanism is institutionally embedded through credit scores, blacklist systems, and certifications for exemplary enterprises. Consequently, structural trust (market mechanisms) and institutional trust (administrative oversight) function in a complementary manner, ultimately translating into psychological trust among consumers. Thus, the three tiers of trust are not independent but are integrated through a chain of relationships: structural → institutional → psychological.

Fifth is the integrative perspective of sustainability. Food safety is no longer merely a hygiene issue but is linked to broad social challenges such as environmental protection, worker protection, and consumer education. Therefore, the Chinese food safety model needs to be restructured as a comprehensive governance system that incorporates ESG and circular economy perspectives. In this regard, sustainability functions as an overarching concept that cuts across the entire three-tier trust model.

However, this model still faces several challenges. First, the opacity of algorithmic governance has the potential to undermine the foundation of structural trust. Second, disparities in institutional capacity among localities compromise the homogeneity of institutional trust. Third, excessive reliance on administrative bodies constrains the flexibility of governance. To address these challenges, improving transparency, establishing accountability, and institutionalizing citizen participation are essential.

8.6 Policy Implications: The Institutionalization of Food Safety as a Public Good and Its Integration through Policy Engineering

8.6.1 Summary of This Book

This book aims to reposition food safety issues in China as a "structure of trust," moving beyond the traditional frameworks of hygiene management and regulatory enforcement. In particular, the book begins by framing food safety within a dual structure of quantitative safety (food security) and qualitative safety (food safety), and clarifies that the information asymmetry and institutional imperfections inherent in the latter are the root causes of consumer distrust.

The book is organized into three parts corresponding to a three-layered model of trust encompassing psychological, institutional, and structural dimensions.

Part I (Psychological Trust) focuses on risk perception and consumer behavior, revealing the existence of a "cognitive gap", the divergence between subjective risk perception and objective safety. Empirical analysis demonstrates that the reliability of information and risk communication are central to trust formation, confirming that food safety issues are closely linked to psychological factors.

Part II (Institutional Trust) analyzes the transformation of food safety legislation and governance, elucidating the process by which governance in China has shifted from a decentralized, reactive model to an integrated, preventive one. Through a comparison of Japan and China, the book examines differences in institutional design and demonstrates that the institutionalization of trust through collaboration among government, businesses, and consumers is indispensable.

Part III (Structural and Sustainable Trust) analyzes the impact of the digital economy and phantomization on food safety, revealing that new supply structures, as exemplified by the OFD market, are transforming risks from "individual" to "structural." At the same time, it demonstrated that traceability and data governance can function as new foundations of trust.

By integrating the analyses from these three parts, this book presents a three-layered integrated model of trust in food safety, comprising psychological trust, institutional trust, and structural trust. These three tiers are not independent; they interact within a chain of relationships, structure → institution → psychology, and determine the stability of trust. Therefore, food safety issues should be understood not merely as matters of regulation or technology, but as governance issues concerning how to integrally design trust across different levels.

Furthermore, this book redefines food safety as a problem of supplying "trust as an institutional public good". Given the information asymmetry in the food market, it is difficult to ensure safety through market mechanisms alone; a trust-supply mechanism that integrates institutions, technology, and the market is indispensable. From this perspective, this book presents the feasibility of constructing a "China food safety model" composed of five elements: (1) state supervision, (2) digital governance, (3) platform collaboration, (4) reputation systems, and (5) sustainability.

Furthermore, this book demonstrates the need to reconnect food safety with the SDGs, particularly Goal 2 (Zero Hunger), Goal 3 (Good Health and Well-being), Goal 12 (Responsible Consumption and Production), and Goal 16 (Peace, Justice and Strong Institutions), and the need to reintegrate it as the foundation of a sustainable society. In this regard, food safety is not merely a discrete policy domain but is positioned as an indicator of institutional quality and societal trust levels.

Overall, the conclusion of this book is clear: food safety is not merely about preventing accidents but is a social process of creating and maintaining trust through the integration of institutions, markets, and technology. Therefore, advancing food safety is nothing less than strengthening governance capacity, that is, improving the "ability to design and deliver trust." The theoretical framework and policy perspectives presented in this book offer important insights not only for China but also for the reconstruction of global food safety governance in the era of the digital economy.

8.6.2 Future Challenges: The Institutionalization of Food Safety as a Public Good and Policy Engineering Integration

Food safety issues have traditionally been addressed primarily within the frameworks of hygiene management and individual regulations; however, the structure of these issues is undergoing significant changes due to the advancement of globalization and digitalization. To respond to these changes, it is essential to adopt the perspective of "policy engineering", which involves the integrated design of problem structures, risk structures, policy instruments, and governing entities, rather than merely piecing together individual policies.

In this book, "policy engineering" refers to an integrated approach to complex social issues that systematically analyzes the interactions among institutions, markets, technology, and actors, and designs optimal policies based on scientific evidence. In other words, policy engineering goes beyond mere policy analysis; it is characterized by treating institutional design itself as the "object of design," while taking into account the interdependencies among different elements.

First, food safety issues are understood as a classic problem of information asymmetry. Food is a trust good whose quality and safety consumers cannot fully assess in advance, and a structural information gap exists between sellers and buyers. As Akerlof (1970) demonstrated, such situations give rise to adverse selection, increasing the likelihood that low-quality goods will remain in the market. In the food market as well, mislabeling and quality deterioration are prone to occur, and the very incompleteness of information becomes a source of risk.

Second, contemporary food safety issues must be understood within the context of the risk society. As Beck (1992) pointed out, risks in modern society are shifting from natural sources to those generated institutionally and artificially. In the context of food safety as well, risks are not accidental incidents but are structurally generated by factors such as the increasing complexity of supply chains, the use of chemicals, international distribution, and digitalization. For this reason, policy responses must shift from ex post regulation to preventive and systematic risk management.

Third, food safety possesses the characteristics of a public good. In light of Samuelson's (1954) argument, while food safety is not a pure public good, it can be understood as a quasi-public good that is not sufficiently supplied by the market due to the existence of information asymmetry. Consequently, government-led institutional design and regulation become indispensable, and food safety should be redefined as an "institutional public good."

Fourth, international standards and risk analysis frameworks form the core of policy engineering integration. The risk analysis framework (risk assessment, risk management, and risk communication) presented by the Codex Alimentarius Commission (2022) enables evidence-based policy design and positions food safety policy as a decision-making process that integrates science and society.

Fifth, digitalization is introducing a new dimension to food safety governance. With the expansion of e-commerce and the platform economy, food distribution is shifting toward a structure dependent on algorithms and data. As the OECD (2016) points out, transparency of information, consumer protection, and enforcement challenges are key issues in digital markets. For this reason, policy engineering must incorporate "data and algorithms" as objects of governance, in addition to traditional regulations and markets.

To summarize, policy engineering in food safety is composed of five theoretical foundations: (1) correcting information asymmetry, (2) responding to a risk society, (3) providing food as a public good, (4) introducing scientific risk analysis, and (5) incorporating digital governance. Within this integrated framework, food safety is redefined as a complex system established through the interaction of institutions, technology, markets, and society.

Therefore, future food safety policies must be constructed as "integrated policy designs" that combine regulation, incentives, information disclosure, technology adoption, and international cooperation, rather than relying on a single policy instrument. Policy engineering provides the theoretical and practical foundation for the sustainable provision of food safety as an institutional public good.

Future research topics include, first, the legal governability of algorithmic oversight; second, an analysis of food safety disparities between regions and between urban and rural areas; third, the correlation between consumer literacy and market quality; fourth, a quantitative assessment of OFD economics and environmental sustainability; and fifth, the theoretical positioning of the Chinese model through international comparison. The integrated analysis of trust, institutions, and structures surrounding food safety presented in this book is not limited to a specific field but can be connected to a general theory of public policy design under uncertainty; the challenge moving forward is to systematize this as "policy engineering."

Overall, the issue of food safety is not merely a matter of public health administration. It is a governance challenge involving the restructuring of the relationship between the state, the market, and society, and a problem of redesigning the order of trust in the digital age.

Through the case of China, this book analyzes food safety from the perspective of policy science (policy engineering) and attempts to connect it to theories of trust and sustainability as institutional public goods. Food safety occupies the forefront of social trust, which is shaped by the accumulation of institutions and practices. As a testing ground for the reconstruction of trust in a digital society, the Chinese case will continue to be an important subject for theoretical and empirical research.

References

Akerlof, G. A. (1970). The market for "lemons": Quality uncertainty and the market mechanism. *The Quarterly Journal of Economics, 84*(3), 488–500. https://doi.org/10.2307/1879431

Beck, U. (1992). *Risk society: Towards a new modernity*. Sage.

Codex Alimentarius Commission. (2022). *General principles of food hygiene: CXC 1-1969*. Food and Agriculture Organization of the United Nations and World Health Organization.

Darby, M. R., & Karni, E. (1973). Free competition and the optimal amount of fraud. *The Journal of Law and Economics, 16*(1), 67–88. https://doi.org/10.1086/466756

Food and Agriculture Organization of the United Nations. (2023). *FAO strategic priorities for food safety within the FAO Strategic Framework 2022–2031*. FAO. https://doi.org/10.4060/cc4040en

Frewer, L. (2004). The public and effective risk communication. *Toxicology Letters, 149*(1-3), 391–397. https://doi.org/10.1016/j.toxlet.2003.12.049

Jaffee, S., Henson, S., Unnevehr, L., Grace, D., & Cassou, E. (2019). *The safe food imperative: Accelerating progress in low- and middle-income countries*. World Bank. https://doi.org/10.1596/978-1-4648-1345-0

Klein, B., & Leffler, K. B. (1981). The role of market forces in assuring contractual performance. *Journal of Political Economy, 89*(4), 615–641. https://doi.org/10.1086/260996

OECD. (2016). OECD Recommendation of the Council on Consumer Protection in E-Commerce. In *OECD eBooks*. https://doi.org/10.1787/9789264255258-en

Pasquale, F. (2015). *The black box society: The secret algorithms that control money and information*. Harvard University Press.

Rochet, J.-C., & Tirole, J. (2003). Platform competition in two-sided markets. *Journal of the European Economic Association, 1*(4), 990–1029. https://doi.org/10.1162/154247603322493212

Samuelson, P. A. (1954). The pure theory of public expenditure. *The Review of Economics and Statistics, 36*(4), 387–389. https://doi.org/10.2307/1925895

United Nations. (2015). *Transforming our world: The 2030 Agenda for Sustainable Development* (A/RES/70/1). United Nations.

United Nations Environment Programme. (2021). *Drowning in plastics: Marine litter and plastic waste vital graphics*. UNEP. https://www.unep.org/resources/report/drowning-plastics-marine-litter-and-plastic-waste-vital-graphics

World Health Organization. (2015). *WHO estimates of the global burden of foodborne diseases: Foodborne disease burden epidemiology reference group 2007-2015*. World Health Organization.

World Health Organization. (2024). *Food safety*. https://www.who.int/news-room/fact-sheets/detail/food-safety

World Health Organization & Food and Agriculture Organization of the United Nations. (2016). *Risk communication applied to food safety: handbook*. World Health Organization. https://iris.who.int/handle/10665/250083

The manufacturer's authorised representative in the EU is Springer Nature Customer Service Centre GmbH, Europaplatz 3, 69115 Heidelberg, Germany. If you have any concerns regarding our products, please contact ProductSafety@springernature.com

Printed and bound by CPI Group (UK) Ltd, Croydon, CR0 4YY
20/10/2026
02236471-0006